MW00714756

McGraw-Hill Data Communications Dictionary

Other McGraw-Hill Communications Books of Interest

AZEVEDO • *ISPF*
BALL • *Cost-Efficient Network Management*
BERSON • *APPC: A Guide to LU6.2*
BLACK • *Network Management Standards*
BLACK • *TCP/IP and Related Protocols*
BLACK • *The V Series Recommendations*
BLACK • *The X Series Recommendations*
CHORAFAS • *The Complete LAN Reference*
COOPER • *Computer and Communications Security*
DAYTON (RANADE, ED.) • *Integrating Digital Services*
DAYTON • *Telecommunications*
FOLTS • *McGraw-Hill's Compilation of Open Systems Standards*
FORTIER • *Handbook of LAN Technology*
GRAY • *Open Systems: A Business Strategy for the 1990s*
HELD, SARCH • *Data Communications: A Comprehensive Approach*
HELDMAN • *Global Telecommunications*
HUGHES • *Data Communications*
INGLIS • *Electronic Communications Handbook*
KESSLER • *ISDN*
KIMBERLEY • *Electronic Data Interchange*
KNIGHTSON • *Standards for Open Systems Interconnection*
LUTZ • *Large Systems Networking with VTAM*
MCCLAIN • *The Open Systems Interconnection Handbook*
MCCLIMANS • *Communications Wiring and Interconnection*
NAUGLE • *Local Area Networking*
NEMZOW • *The Ethernet Management Guide*
OWEN • *Digital Transmission Systems*
RADICATI • *Electronic Mail*
RANADE • *Advanced SNA Networking*
RANADE • *Introduction to SNA Networking*
RHEE • *Error Correction Coding Theory*
ROHDE • *Communications Receivers*
RORABAUGH • *Communications Formulas and Algorithms*
ROSE • *Programmer's Guide to Netware*
SARCH • *Integrating Voice and Data*
SARCH, ABBATIELLO • *Telecommunications and Data Communications Fact Book*
SCHLAR • *Inside X.25*
TUGAL • *Data Transmission*
UNGARO • *Networking Software*
WATERS • *Computer Communications Networks*
WHITE • *Interworking and Addressing*

McGraw-Hill Data Communications Dictionary

Definitions and Descriptions of General and SNA Terms, Recommendations, Standards, Interchange Codes, IBM Communications Products, and Units of Measure

William F. Potts

Uyless Black
Consulting Editor

McGraw-Hill, Inc.

New York St. Louis San Francisco Auckland Bogotá Caracas Lisbon London Madrid Mexico Milan Montreal New Delhi Paris San Juan São Paulo Singapore Sydney Tokyo Toronto

Library of Congress Cataloging-in-Publication Data

Potts, William F.
McGraw-Hill data communications dictionary : definitions and descriptions of general and SNA terms, recommendations, standards, interchange codes, communications products, and units of measure / William F. Potts.
p. cm.
Includes bibliographical references (p.) and index.
ISBN 0-07-003154-1
1. Data transmission systems—Dictionaries. 2. Computer networks--Dictionaries. I. McGraw-Hill, Inc. II. Title. III. Title: Data communications dictionary.
TK5105.P675 1992
004.6'03—dc20

91-48333
CIP

1 2 3 4 5 6 7 8 9 0 DOC/DOC 9 8 7 6 5 4 3 2

ISBN 0-07-003154-1

The sponsoring editor for this book was Neil Levine, the editing supervisor was Joseph Bertuna, and the production supervisor was Pamela Pelton. It was set in New Century Schoolbook.

Printed and bound by R. R. Donnelley & Sons Company.

To John W. King III, wherever you may be

Contents

Acknowledgments

In addition to the books listed in the Bibliography, the author wishes to acknowledge the following invaluable sources of information:

ANSI (American National Standards Institute).
CCITT (The International Telegraph and Telephone Consultative Committee—part of ITU).
EIA (Electronic Industries Association).
Global Engineering Documents, Irvine, CA.
IBM Corporation.
IEEE (Institute of Electrical and Electronic Engineers).
ISO (International Organization for Standardization).
ITU (International Telecommunications Union).

Trademarks

Preface

Definitions of data communications terms take many forms, of which the most common is the tersely written appendix to be found in many data communications text books.

Vendors of data communications equipment and software also provide glossaries, of which the best-written and most accurate are produced by large, well-established companies like IBM. Inevitably, they are biased in favor of the products with which they are associated.

The goal of this book is to provide comprehensive definitions, while avoiding the lengthy, detailed, subject-by-subject discourse that is the domain of the textbook.

How this book is organized

This book has six sections.

Section 1 deals with General, SNA, and Vendor Terms. General terms may be fundamental (e.g., full duplex), arbitrary (e.g., Network layer, which is defined as a component of the OSI Model, itself an arbitrary concept), or any of a variety of things such as units of measure (e.g., bit/s). Vendor terms are special terms, descriptive of products, capabilities, protocols, architectures, etc., offered by various vendors, including IBM. SNA terms are a special case of vendor terms, and they reflect the dominance of IBM and SNA in host-centered networks.

Section 2 deals with Recommendations and Standards. It lists the ANSI, EIA, IEEE, and ISO data communication standards, plus the CCITT I-Series, V-Series, and X-Series Recommendations. Where appropriate, it provides an abstract or, in some cases, a fairly detailed treatment. For those involved in the creation, acquisition, or documentation of products which must conform to one or more of the standards or Recommendations, this section provides a starting point. Those requiring a detailed knowledge must, of course, refer to the documentation of the relevant standards organizations. Their names and addresses, plus those of resellers of standards documentation, are provided at the end of the section.

Sections 3 and 4 deal with IBM Data Communications Products. Section 4 includes 3270 and 5250 terminology and information, in tabular form, on the displays and printers associated with the IBM 3270 and 5250 Information Display Systems (IDS), both of which are key SNA subsystems. Section 3 describes all of the other significant IBM data communications products, including controllers for information display systems.

Section 5 contains tables of the two most important data communications interchange codes, namely CCITT International Alphabet No. 5 (also known as ISO 646 and, in its U.S. implementation, as both ANSI X3.4 and ASCII) and EBCDIC, and includes control-code descriptions.

Finally, **Section 6** deals with an important standards issue, namely the means defined by the *Système International (SI)* for expressing units of measure. A number of governments now require that prospective vendors adhere to SI conventions in all proposal specifications. Although SI documentation does not deal specifically with data communications, it does provide firm guidance.

Notation

For SNA definitions involving descriptions of bytes, this book uses IBM's customary notation. For definitions involving descriptions of octets or bit strings (and, typically, based on CCITT Recommendations or related standards), the notation used in CCITT documents is used.

IBM notation, within a byte, uses ascending bit numbers, starting with the most significant bit (MSB), thus:

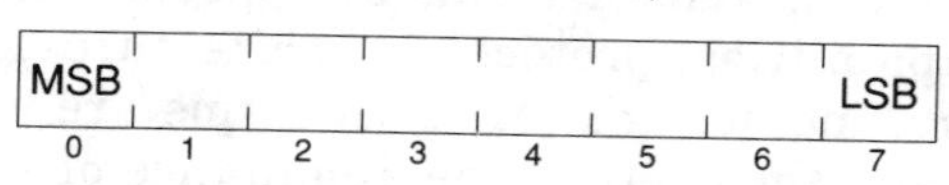

CCITT notation also uses ascending bit numbers, but starts with the least significant bit (LSB) of each octet and, where fields span two or more octets, does not reset at the octet boundary, thus:

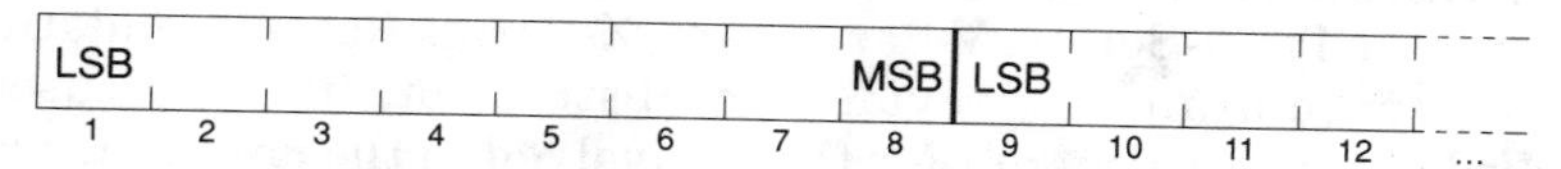

Although the placement of the least significant bit at the beginning violates traditional numbering conventions, it corresponds to the sequence in which bits are transmitted over a communication link.

The reader should also be aware of a third notation, typically used in personal computer documentation, in which numbering within byte is from most to least significant bit, but with descending bit numbers, thus:

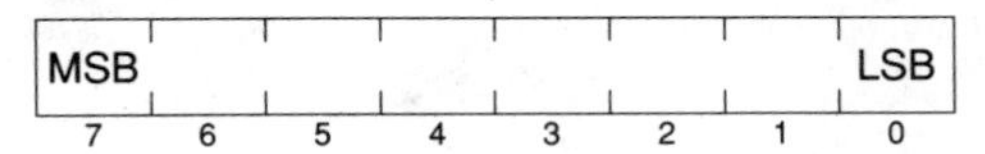

A Note on Spelling and Idioms

The field of data communications is indisputably international. This book is, accordingly, aimed at an international readership. As far as possible, it is written in "international English" — idiom-free and with American spelling. Where American spelling offers a choice, this book uses the more universally accepted option (e.g., *analogue*, not *analog*; *dialling*, not *dialing*).

McGraw-Hill Data Communications Dictionary

Section

1

General, SNA, and Vendor Terms

access method A software component of a computer operating system which controls the flow of data between application programs and either local or remote peripheral devices.

ACF/NCP Advanced Communications Function/Network Control Program. Software, resident in an IBM communication controller (3705, 3720, 3725, or 3745), which controls communication between the host system (or systems) to which the controller is attached and other devices on the network (usually, but not necessarily, an SNA network). For a communication controller not attached to a host system, ACF/NCP can provide line concentration functions.

ACF/TCAM Advanced Communications Function/Telecommunications Access Method. The current version (Version 3, introduced in March 1985) is a subsystem which operates as an ACF/VTAM application program. It includes a message handler, queuing capabilities, and a DCB (data control block) application program interface.

ACF/VTAM Advanced Communications Function/Virtual Telecommunications Access Method. System program which runs on IBM host computers and controls communication between host application or service programs and terminals. The prefix *ACF* distinguishes contemporary VTAM releases (which support multidomain networks) from earlier ones which provided much less comprehensive support.

ACF/VTAME Advanced Communications Function/Virtual Telecommunications Access Method—Entry. Obsolete version of ACF/VTAM used in conjunction with VSE on smaller IBM mainframes with built-in communications adapters. It was functionally similar to the combination of ACF/VTAM and ACF/NCP. Current releases of ACF/VTAM include the functions of ACF/VTAME.

acoustic coupler A low-speed anisochronous modem, designed to transmit and receive the analogue signals in audible form via a telephone handset. In the United States, Canada, Britain, and elsewhere, the introduction of modular telephone jacks and the removal of restrictions on the direct connection of approved devices to telephone circuits have eliminated the need for acoustic couplers.

ACTLU See *SSCP-LU,* under *session.*

ACTPU See *SSCP-PU,* under *session.*

adaptive equalization See *equalization.*

adaptive routing Network routing scheme that can adapt to changes in traffic patterns, congestion, failures, etc.

ADCCP Advanced Data Communication Control Procedures. A bit-oriented data link protocol. It is defined as ANSI Standard X3.66 and also as U.S. Federal Standard 1003. It is identical to HDLC (ISO 3309:1984 and ISO 4335:1987).

address 1. In SNA, a 16-bit value which uniquely identifies a subarea and an element (e.g., LU) in the network. Each NAU (LU, PU or SSCP) has its own address. Subarea addresses are associated with communication controllers.

2. A value, used to uniquely identify a secondary station on a multipoint data link.

address field See *HDLC frame* and *SDLC frame.*

ADPCM Adaptive differential pulse-code modulation. Variant of pulse-code modulation in which amplitudes are represented by 4-bit, rather than 8-bit, values. The data rate is similarly reduced—from 64 kbit/s to 32 kbit/s. See also *PCM.*

AlphaWindow Proposed standard for text-mode ASCII display terminals, providing for multiple, windowed communication sessions. Requires host routing software.

amplitude modulation (AM) Modulation technique in which the amplitude of the signal (e.g., continuous tone) is set to one of a number (typically 2) of discrete levels to represent the values 0 and 1 (or the values of bit combinations). Often combined with other modulation techniques. See *QAM*.

analog Alternative U.S. spelling of analogue.

analogue Something that bears an analogy (correspondence in some respects between things otherwise dissimilar) to something else. In data communications, representation of digital values by means of variations in waveforms.

analogue loopback See *loopback test*.

anisochronous Descriptive of the operation of two or more devices (e.g., modems) not governed by a common clocking mechanism. Anisochronous modems are commonly referred to as asynchronous modems. Antonym of isochronous. See *isochronous*.

ANSI American National Standards Institute. ANSI, which is responsible for the establishment of many standards, including a number of data communications and terminal standards, is a recognized representative body within CCITT. For specific standards, see *ANSI Data Communications Standards*, starting on p. 80.

answer tone Tone signal, with a frequency between 2025 and 2225 Hz and a duration of at least 1.5 s, used by an answering modem to indicate its ready condition to an originating modem.

API Application Program Interface. A means by which an application program gains access to system resources, often for the purpose of communication (the sending and receiving of data). In the specific area of terminal emulation, an API provides for the simulation of keystrokes and for writing into and reading from the presentation space (device buffer). With 3270 terminal emulation, it may also provide for the sending and receiving of structured fields.

APPC Advanced Program-to-Program Communication. A capability, involving Logical Unit Type 6.2 and its associated protocols, which allows communication between processes (e.g., application programs) in an SNA or APPN network without the involvement of a common host system or of terminal emulation.

Application layer See *OSI*.

APPN Advanced Peer-to-Peer Networking. IBM networking architecture in which communication between end points (e.g., users and resources or a pair of users) is at a peer level and not dependent on a central host-system control point. Information on network topology is distributed among Network Nodes, each of which maintains information on all links between Network Nodes, plus its own End Nodes and the users they serve.

APPN was first introduced as a proprietary networking scheme. However, in March 1991, IBM opened it up to developers and other vendors as an architecture, defined as an extension to SNA and SAA. IBM's initial open architecture documentation provides End Node specifications. Network Node specifications are expected to follow. End Nodes are classed as SNA physical units (PU Type 2.1), supporting LU Type 6.2 for end users and resources.

Also in March 1991, Apple Computer, Inc., Novell, Inc., Systems Strategies, Inc., and Siemens/Nixdorf Informationssysteme AG announced their intention to provide end-node products. See also *session*.

APS See *DIA/DCA*.

ARPANET Advanced Research Projects Agency (ARPA) Network. Packet-switching network, developed in the late 1960s to link government computing facilities. ARPANET is now part of the Internet, and ARPA has since been renamed DARPA (Defense Advanced Research Projects Agency). See also *Internet* and *TCP/IP*.

ARQ Automatic request for repeat. Type of error control used in most synchronous data link protocols (e.g., SDLC, HDLC, BSC) in which a receiving station automatically requests a retransmission from the transmitting station when an error is detected. See also *FEC*.

ASCII American National Standard Code for Information Interchange. Seven-bit code, intended as a U.S. standard (ANSI X3.4) for the interchange of information among communications devices. The first 32 values are reserved for control codes, with the remaining 96 available for graphics (alphabetic, numeric, and special characters). Several 8-bit variants exist, including IBM's PC Extended ASCII character set, in which even the control codes have graphic assignments. Other specific variants include Hewlett-Packard's Roman-8 and ECMA-94 (European Computer Manufacturers' Association). See also *ISO 646, in Sec. 2,* and *CCITT International Alphabet No. 5,* in *Sec. 5*.

ASCII host Informal term for a host computer, often but not necessarily non-IBM, supporting communication in start-stop mode with ASCII terminals.

ASCII terminal Informal term for a start-stop terminal supporting one of the variants of CCITT International Alphabet No. 5 (e.g., ANSI X3.4 (ASCII), British Standard Data Code, etc.). Typically, an ASCII terminal is a keyboard/display device, sometimes with an attached auxiliary printer.

asserted circuit Interchange circuit which has been placed in an ON condition by one of the connected devices (e.g., DTE) to indicate a status or a requirement to the other device (e.g., DCE). For example, when a DTE requires control of a data link in order to transmit, it asserts the Request-to-Send (RTS) circuit. In normal practice (CCITT Recommendation V.28, EIA-232-D, etc.), control circuits are asserted (turned ON) by the application of a positive voltage (space condition), with a negative voltage (mark condition) denoting the nonasserted (OFF) state.

async Abbreviation of asynchronous.

asynchronous transmission See *start-stop*.

audible alarm Device used to call a terminal or PC operator's attention to some event or condition. On ASCII terminals, the BEL code (07) is used to trigger the audible alarm.

audio response unit Device which can compose spoken messages from analogue or digitized recordings of syllables, numerals, etc. Typically used to respond to inquiries made with DTMF-equipped telephones. Larger audio response units can transmit many messages concurrently. See also *DTMF*.

auto answer Combined DCE/DTE capability providing for the automatic answering of incoming calls and, subject (usually) to the verification of user identity, password, and other parameters, the establishment of a communications session.

autokey A capability which allows keystrokes to be recorded in a special file and replayed on request. Used to avoid repetition of commonly used keystroke sequences.
See also *record/play/pause* in *Sec. 4*.

Automatic Calling Unit (ACU) See *Bell modems, Hayes-compatible modems,* and *CCITT Recommendations V.25* and *V.25 bis (Sec. 2)*.

automatic polling IBM term for a communication controller capability in which terminals are polled on the basis of address entries in a list (*polling list*) supplied by the attached host computer. When the communication controller's polling software reaches the end of the list, it starts over at the beginning. The automatic polling process may involve the deletion of address entries corresponding to nonresponding terminals, with such entries being reinstated every so often to allow for the possibility that they have subsequently been activated. Also called *autopoll*. See also *polling*.

backbone network Facility, usually high-speed, to which subnetworks (particularly LANs) are attached. Provides for communication among the subnetworks and between subnetworks and shared services, such as file servers, database servers, gateways, host systems, etc.

backward channel Means by which a DCE, while nominally in receive mode, can signal the transmitting DCE. May be limited to on/off signalling or may be capable of limited speed data transfer.

balanced circuit Circuit (in a DTE-to-DCE interface, for example) which uses two conductors, rather than a common return circuit (signal ground). Unlike unbalanced circuits, balanced circuits are not sensitive to differences in ground potential. See also *CCITT Recommendation V.11* in *Sec. 2*.

balun Balanced-to-unbalanced connector. Used to connect twisted-pair cable to a coaxial port on a controller, terminal, or similar device. See also *coaxial cable*.

band A range of frequencies.

bandpass filter Circuit, within an analogue device, which allows the passage of a limited range of frequencies (i.e., it filters out frequencies above and below the required range).

bandwidth The difference, expressed in hertz (Hz), between the two limiting (upper and lower) frequencies of a band.

baseband signalling Form of transmission which uses discrete pulses, without modulation. Often called *baseband transmission*.

batch mode Operational mode in which input to or output from a process is transmitted as a single set of successive messages. Contrast with *interactive mode*. See also *JES*.

baud Signalling rate unit for analogue communications. One baud is equal to one change of state per second. Where there are two possible states (e.g., two tone frequencies), 1 baud equals 1 bit/s. Where there are 2^n possible states (e.g., four possible phase shifts in a sinusoidal wave), 1 baud equals n bit/s. In low-speed transmission, where modems with frequency-shift keying are used, the terms *baud* and *bit/s* may be used interchangeably. Otherwise the two terms are never interchangeable.

BCC Block Check Character. Character at the end of a transmission block, whose value is calculated in some way from the value of all data characters in the block. When the block is received, an identical calculation is performed and the result is compared with the transmitted BCC. The calculation may be the logical addition (no carries) of the binary values of all characters or may be based on a polynomial or other scheme. Many file transfer protocols (e.g., X-modem) use a single BCC, often referred to as a checksum. Most synchronous communication protocols use two block check characters. See *CRC* and *FCS*.

BCD Binary Coded Decimal. A code defined with 6 bits per character and capable, therefore, of representing up to 64 unique values, each representing one of the 26 letters of the English alphabet (upper case only), a digit (0 to 9), or a special or punctuation character. BCD is no longer used. See *EBCDIC*.

BCDIC Binary Coded Decimal Interchange Code. An extension of BCD, used in data communications, in which an upshift and a downshift character are defined, thus allowing subsequent alphabetic code values to represent either upper- or lowercase letters and allowing the other values to represent alternative characters. BCDIC is no longer used. See *EBCDIC*.

BCH code Forward error-correcting (FEC) code, named for its developers, Bose, Chaudhouri, and Hocquenham, and useful for error correction in short messages. See also *FEC*.

BDLC Burroughs Data Link Control. Data link protocol, based on but not compatible with HDLC.

Begin Chain See *chaining protocol*.

beginning flag field See *flag field*.

Bell modems Prior to FCC rulings eliminating the requirement for data access arrangements (DAAs) between non-telephone-company modems and telephone circuits, the Bell Telephone Company dominated the U.S. and Canadian modem market, offering modems with internal DAAs. Bell offered six major series of modems, plus a companion series of automatic calling units. For a long time, the large intalled base of Bell modems forced competing modem companies to produce modems that were compatible with one or more of Bell's de facto standards. Many manufacturers also offered their own standard, either as a setup option or, for an answering modem, as a switchable option. Even though several other de facto standards have been introduced by some of the larger modem manufacturers (Hayes, Microcom, Telebit, etc.), and conformance with CCITT V-series modem recommendations (especially V.32, V.42 and the newer V.42 *bis*) is becoming widespread in the U.S. and Canada, the Bell standards still exist and are, therefore, important. Several CCITT modem recommendations (V.21, V.22, V.26, V.27, V.29) are variants (although not compatible variants) of Bell standards.

103/113 series A family of full-duplex, anisochronous modems, operating at rates up to 300 bit/s, using frequency-shift keying (FSK). Offered in manual-originate-only, auto-answer-only, and manual-originate/auto-answer versions. An originating modem transmits on a center frequency of 1170 Hz, with mark and space frequencies of 1270 and 1070 Hz, respectively. An answering modem transmits on a center frequency of 2125 Hz, with mark and space frequencies of 2225 and 2025 Hz, respectively.

212 series A family of full-duplex modems, operating isochronously at 1200 bit/s and, in some models, anisochronously at rates up to 300 bit/s. Uses dibit phase-shift keying at 1200 bit/s and frequency-shift keying at up to 300 bit/s.

201 series A family of isochronous modems, operating at 2000 or 2400 bit/s, in half-duplex mode on two-wire circuits and in full-duplex mode on four-wire circuits. Uses dibit phase-shift keying with 0°, 90°, 180°, and 270° phase shifts. May, alternatively, use 45°, 135°, 225°, and 315° phase shifts. The 201A operates at 2000 bit/s. The 201B and 201C operate at 2400 bit/s.

202 series A family of anisochronous modems operating at rates up to 1200 bit/s on dial-up lines (202C and 202S) and up to 1800 bit/s (marginally) on C2-conditioned leased (private) lines (202D and 202T), using frequency-shift keying (FSK). Operation is half duplex on dial-up or leased two-wire circuits, full duplex on four-wire circuits. Transmission uses a center frequency of 1700 Hz, with mark and space frequencies of 1200 and 2200 Hz, respectively. In two-wire operation, a 387-Hz secondary (backward) channel may be used for on/off signalling only at a maximum switching rate of 5 baud. Some 202-compatible modems (e.g., Universal Data Systems) provide a full reverse channel operating at up to 150 bit/s.

208 series A family of isochronous modems, operating at 4800 bit/s, in half-duplex mode on two-wire circuits and in full-duplex mode on four-wire circuits. Uses tribit phase-shift keying to modulate an 1800 Hz signal, with phase shifts of 22.5°, 67.5°, 112.5°, 157.5°, 202.5°, 247.5°, 292.5°, and 337.5° for tribit values (after scrambling) of 000, 001, 010, 011, 100, 101, 110, and 111, respectively.

209 series A family of isochronous modems, operating at 9600 bit/s in full-duplex mode on four-wire circuits with D1 conditioning. Can operate at fallback rates of 7200 or 4800 bit/s.

Uses QAM (quadrature amplitude modulation) to modulate a 1650-Hz signal. Each change of state is used to represent 4 consecutive bits (quadbit, or tetrabit), derived by scrambling the data stream. Handles envelope delay distortion and attenuation distortion with an automatic adaptive equalizer.

A time-division multiplexing option, in conjunction with up to four DCE/DTE interface connections, provides for the following data stream combinations:

- One at 7200 bit/s and one at 2400 bit/s
- Two at 4800 bit/s
- One at 4800 bit/s and either one or two at 2400 bit/s
- Up to four at 2400 bit/s

801 series Automatic Calling Units (ACUs) Devices incorporating the parallel automatic calling interface defined by the EIA-366-A standard. The ACU's relationship to the DTE and to the modem are illustrated in the following diagram. There were two basic models of this now obsolete device. The 801A supported pulse dialling at 10 pulses per second. The 801C supported tone (DTMF) dialling at 7.7 digits per second.

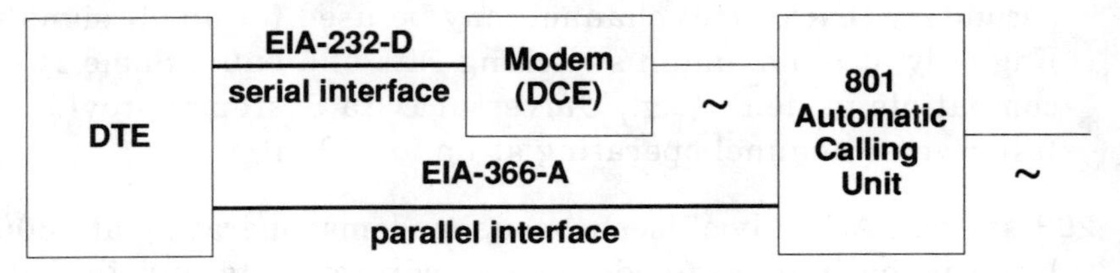

Binary Synchronous Communication See *BSC*.

BIND Command sent from one NAU to another (e.g., a primary LU to a secondary LU) to activate a session between them. See also *session* and *Logical Unit*.

binding The process, in SNA, by which a session is established between network-addressable units (NAUs). See also *session* and *Logical Unit*.

bipolar Having both positive and negative voltages.

bisync Informal abbreviation for Binary Synchronous Communications. See *BSC*.

bit A binary digit, capable of having only two values, 0 and 1.

bit error rate (BER) For a standard error-testing pattern, the ratio of bits in error to total bits received.

bit error rate test (BERT) A test in which a pseudo-random standard pattern, typically 63, 511, or 2047 bits in length, is transmitted over a communications channel. The resulting bit error rate (defined above) indicates the degree of line impairment. Testing methods and test patterns are defined by CCITT Recommendation O.153.

bit rate Informal term for data rate, when it is measured in bits per second.

bit stuffing See *zero-bit insertion*.

bit/s Bits per second. Unit of measure for data transmission. Often confused with baud. See *baud*.

BIU Basic Information Unit. In SNA, a unit of information consisting of an RU (Request/Response Unit) and an RH (Request/Response Header) added by the Transmission Control layer. See also *BTU, PIU, RH, RU,* and *TH*.

.... Basic Information Unit (BIU)	
RH	Request/Response Unit (RU)

blink A display attribute in which the displayed information is repeatedly turned off and back on, typically at intervals of about 0.5 s.

Block Check Character See *BCC*.

block mode Operating mode of a start-stop terminal, usually involving half-duplex operation and characterized by the transmission of blocks of information consisting of single or multiple entry fields. In many cases, a transmitted block consists of a single line of text. Contrast with *conversational mode*.

block number 12-bit value which identifies a specific class within an SNA physical unit type. For example, within PU Type 2, a block number of 017 (hex) identifies an IBM 3174 establishment controller, 3274 cluster controller, or compatible equivalent.

bps See *bit/s*. See also *Sec. 6*.

bracket In SNA, an arbitrary grouping of related normal-flow chains. Used when multiple request units are associated with a transaction or a work unit. A work unit might, for example, be the data stream corresponding to a print file sent from a host system to a remote printer.

The first request unit in the first chain of a bracket is marked with a Begin Bracket Indicator. The first request unit in the last chain is marked with an End Bracket Indicator. See also *chain*.

break A signal sent over a backward (secondary) channel by a receiving start-stop terminal on a half-duplex circuit (or, in some cases, over the transmit channel on a full-duplex circuit), usually to indicate a requirement to transmit. Break was originally a teletypewriter term, designating a key used by a receiving operator to interrupt the current loop between a pair of terminals long enough to indicate to the sending operator a desire to send a message in the other direction.

CCITT Recommendation X.28 includes a very specific definition of a break signal, to be used between a start-stop mode DTE and a PAD (packet assembler/disassembler). It is used to signal to the PAD without loss of transparency, and may also be used for signalling from the PAD to the DTE. The break signal is defined as the transmission of binary 0 for more than 135 ms. It must be separated from any following start-stop character or other break signal by the transmission of binary 1 for at least 100 ms.

breakout box A device for monitoring and manipulating the interface signals between a DTE and a DCE (or between a pair of DTEs).

A typical EIA-232-D (CCITT V.24/V.28) breakout box has two DB-25 connectors (see *ISO 2110,* in *Sec. 2*), LEDs (light-emitting diodes) for the more common data, control and timing circuits, DIP switches for all conductors, and 25 jumper-wire sockets on each of the DTE and DCE sides. Often the LEDs are in red/green pairs, with red representing a positive voltage (0, or space condition) and green representing a negative voltage (1, or mark condition). Voltages in the range -3 V to +3 V will not light the LEDs, and rapidly alternating polarity (on data or timing circuits) will light both LEDs. Sometimes, tristate LEDs are used, with red for positive, green for negative, and orange for alternating. The user may enable or disable any circuit by means of the DIP switches or may use short jumper wires with the sockets to correct cable wiring errors or, for example, to create a null modem. Additional LEDs are usually provided, allowing the user to include them in any circuit by means of the jumper wires. Finally, a battery and two additional jumper-wire sockets are provided, one as a source of +12 V and the other as a source of -12 V (relative to ground return), allowing the user to force mark or space condition on any circuit. Additional features may include "pulse traps," used to capture signals too short in duration to illuminate an LED. See also *null modem*.

breakout panel A control or "patch" panel, typically rack-mounted, which serves the same purpose as a breakout box.

BRI Basic Rate Interface. See *ISDN*.

bridge A device used singly, or with other like devices, to connect two or more similar local area networks (LANs) in order to allow workstations on one network to communicate with resources on the other network or networks as though they resided on the same network. A single-unit bridge is used where LANs are in the same location. Multiple bridges (one on each LAN) are used where the interconnected LANs are some distance apart.

broadband 1. Descriptive of a circuit or a facility capable of carrying signals of greater bandwidth than can be supported by voice-grade circuits. In this context, the term "wideband" is often preferred.

2. Descriptive of channels accommodated, in large numbers, on a transmission medium such as coaxial cable, where each channel is modulated by a different radio-frequency carrier, typically in the 50- to 500-MHz range.

3. For ISDN, a service or system with channels that can support rates higher than the primary rate (1.544 or 2.048 Mbit/s).

brouter Device combining the characteristics of a bridge and a router. See also *bridge* and *router*.

BSC Binary Synchronous Communication. A link-level data communications protocol, defined by IBM, in which transmission is synchronized by block (message) and in which the receiving device must acknowledge the error-free receipt of successive blocks of data with alternating acknowledgment messages, arbitrarily designated as ACK0 and ACK1. Two successive acknowledgments of the same kind (e.g., two ACK0's) are an indication of a lost block. The BSC protocol is most commonly used for the transmission of EBCDIC data but may also be used for ASCII data or Six-bit Transcode.

For a description of the method of synchronization used for BSC, see *synchronous*. For a full description of BSC, please refer to IBM's BSC documentation.

BTU Basic Transmission Unit. In SNA, a unit of information consisting of one or more PIUs (Path Information Units). The information field of an SDLC I frame (information frame) is a BTU. See also *BIU, PIU, RH, RU,* and *TH*.

.... Basic Transmission Unit (BTU)		
PIU 1	- -	PIU n

buffer 1. Memory allocated for the temporary storage of incoming and/or outgoing data.

2. Memory containing data for presentation to a user (by way of a video display, printer, or similar device). See also *presentation space*.

burst One of a series of successions of bits, frames, or other elements of data, occurring at irregular intervals.

bus A common connection among a number of devices. Use of the bus is controlled by an arbitration scheme, which may be based on contention, token possession, or time-slice allocation.

byte A group of 8 bits, representing any of 256 values. A byte may represent a single binary number, 8 bits of a longer binary number, two decimal digits, one decimal digit with plus or minus sign, or a graphical character (e.g., a letter, a number or any symbol). The term byte was introduced by IBM, in 1964, with the introduction of the System/360 series of mainframe computers. The generic term, used by CCITT and similar organizations, is *octet*.

C conditioning See *line conditioning*.

carrier See *modem carrier*.

carrier frequency See *modem carrier*.

CCITT Comité Consultatif International Télégraphique et Téléphonique (Consultative Committee, International Telegraph and Telephone). See *Sec. 2*.

CCS Common Communications Support. See *SAA*.

cell relay Proposed circuit-switched service offering, intended for use over very-high-speed connections.

CEPT Conference of European Postal and Telecommunications administrations. Organization responsible for the establishment of common operating and interworking arrangements and practices among the European PTTs. See also *PTT*.

chain In SNA, a group of request units that is treated as a single entity for the purpose of error recovery. If any request unit in a chain is in error, the entire chain is discarded. Chaining is handled by SNA's Data Flow Control layer. See also *chaining protocol*.

chaining protocol In SNA, a method of logically defining and transmitting a complete unit of data, whether it is contained in a single RU or in two or more RUs. The Request/Response Header (RH) of each RU can contain Begin Chain (BC) and/or End Chain (EC) indicators. RUs can be First in Chain (FIC—BC indicator only), Middle in Chain (MIC—neither indicator), Last in Chain (LIC—EC indicator only), or Only in Chain (OIC—both indicators). Valid chains are FIC, LIC; FIC, MIC, ..., LIC; and OIC. Any other sequence causes a chaining error. See also *chain*.

channel 1. A means of one-way transmission.

2. On a large computer (e.g., IBM mainframe system), a high-speed device used for communication between the central computer and peripheral devices, such as direct-access storage subsystems, tape subsystems, printers, communication control-

lers, and local cluster controllers. Channels contain processor logic, used for the execution of channel programs. Channel programs are initiated by the host system, which is notified by means of an interrupt when they have completed their task. There are three basic IBM mainframe channel types, distinguished by the way they schedule the transfer of data:

block multiplexer By means of a process called "disconnected command chaining," can handle the execution of multiple, concurrent channel programs. However, data can be transferred (at very high speed) to or from only one device at a time. During data transfer, channel program execution is temporarily suspended. Overlapping operations are possible during the waiting time for the completion of certain peripheral device functions (e.g., seek time and rotational delay time on disk drives). Communication controllers, such as the IBM 3745, are normally attached to block multiplexer channels, as are local SNA Cluster/Establishment Controllers (3274 and 3174). Other attachable devices include disk and tape subsystem controllers.

byte multiplexer Can handle multiple, concurrent channel programs and multiple, concurrent (interleaved) data transfers to and from peripheral devices requiring relatively modest data transfer rates. Each channel program operates within an addressable subchannel. Multiple subchannel addresses may be assigned to a single peripheral controller (e.g., non-programmable communication controller, or non-SNA local cluster controller), requiring that the host software address every controller-attached device (e.g., communication line, 3270 display) individually. For higher-speed devices, a byte multiplexer channel also has "burst mode" operation.

selector An obsolete channel type, capable of executing only one channel program at a time.

See also *ESCON Architecture Facility.*

channel anomalies See *distortion* and *noise*.

character 1. A graphic symbol, such as a letter, a numeric digit, punctuation, or other special symbol.

2. A group of n bits, usually representing one of 2^n possible graphic symbols and/or control functions (control characters). The value of n may be 8 (EBCDIC and extended ASCII), 7 (ASCII), 6 (BCD, BCDIC), 5 (Baudot (telegraph) code), 16 (for ideographic characters, such as Chinese, Kanji, etc.), or any

equipment-dependent number. Where shift in and shift out (or upshift and downshift) control functions are defined, the total number of symbols and control functions may exceed 2^n.

CICS/VS Customer Information Control System/Virtual Storage. IBM host teleprocessing (data communications) monitor which provides an environment for transaction-oriented terminal applications.

circuit A means of communicating between two points, consisting of send and receive channels.

circuit switching Call establishment mechanism providing end-to-end clear-channel temporary circuits.

clear-channel Permitting direct end-to-end communication. Commonly used to refer to circuit-switched or permanent-circuit ISDN connections. Also applicable to conventional circuit-switched or leased (private) line connections.

clipping The loss of 1 or more bits at the beginning of a transmission, typically caused by a delay in line turnaround or echo suppression. (May also occur in voice communication, with the loss of the beginning of an initial syllable.)

cluster controller See *Cluster Controller* and *Establishment Controller* in *Sec. 4*.

coax Common abbreviation for coaxial cable.

coaxial Having a common axis.

coaxial cable Cable consisting of a center conductor (usually copper), surrounded by a braided tubular conductor of uniform diameter. The two conductors are separated by a tubular polyethylene insulator and the cable is encased in a vinyl sheath. See also *coaxial cable, Category A,* and *Category B* in *Sec. 4*.

code A system of rules and conventions, according to which the signals representing data can be formed, transmitted, received and processed. See also *character, ASCII, EBCDIC,* and *interchange code*.

code page IBM term for a specific assignment of graphic characters and control functions to all code points within a code set. IBM has defined a number of national-language code pages, each with an identifying number. Loadable code pages provide for instant national-language or special-use adaptation of displays and printers.

code point A specific value of a code (an n-bit value, out of 2^n possible values). A single code point may represent a control function or one (or more than one) graphic symbol. Variations may exist because of differing national-language implementations or because of the effect of preceding shift in and shift out characters. See also *character.*

code set All possible values of a code. For an n-bit code, the code set consists of 2^n values. For example, EBCDIC, which is an 8-bit code, has 256 (2^8) possible values. The number of characters a code set can represent may exceed the total number of possible unique values. See also *character, shift in,* and *shift out.*

communication controller Device, usually attached to a channel on an IBM mainframe computer, used to control a number of communications lines. Speed, communications protocol and type of connection may vary from line to line. See *IBM 37x5* in *Sec. 3.*

communication controller node General term for SNA Physical Unit Type 4. See also *Physical Unit.*

communication link See *data link.*

communications adapter Component within a DTE (e.g., communication controller, personal computer, etc.), containing the physical communication (DTE-DCE) interface, the logic to control that interface, and commonly, the logic to handle the data link control protocol.

compandor Compressor/expandor. A device used in conjunction with voice-grade channels which compresses the dynamic range of the signal strength of transmitted signals and expands the dynamic range of received signals. In practice, the compression process reduces the signal strength slightly at the high end and increases it considerably at the low end, with the goal of raising the low end above the level of any noise on the channel. Compandors at either end of a channel must be perfectly matched, or they will produce compandor distortion. See also *distortion.*

constant-ratio code A code in which all code points have the same ratio of 1s to 0s. Since any modification of a bit will violate the constant ratio, error checking is implicit. See also *four-of-eight code.*

control character Within a code set, a character intended to initiate, modify, or stop a control function. *Sec. 5, Interchange Codes,* describes control characters for ISO 646 (ASCII) and EBCDIC.

control field Eight-bit or 16-bit field, immediately following the address field in an HDLC or SDLC frame, used to define the frame type and to control the sequence of frames. See *HDLC frame* and *SDLC frame.*

control function See *control operation.*

control operation An action affecting the tranmission, formatting, processing, interpretation, or recording of data.

controller See *cluster controller, establishment controller (Sec. 4),* and *communication controller.*

conversational mode 1. Operating mode of a start-stop terminal, characterized by full-duplex operation and the echoing, by the remote DTE, of all keystrokes entered at the terminal. Contrast with *block mode.*

2. Alternative term for *interactive mode* (see also).

cooperative processing Processing involving an application, portions of which are resident on separate devices, often with inter-device communication over data links.

CP-CP session See *session.*

CRC Cyclic Redundancy Check. A means of ensuring the integrity of a transmitted data block in which a 16-bit sum is appended to the end of the block by the transmitting device. The sum is generated from cyclically weighted values, each of which is added to the sum (binary addition, with no carry) if it corresponds to the occurrence of a 1 bit. The receiving device uses the same process to generate a comparison sum, which should be equal to the transmitted sum.

crosstalk Interference from an adjacent analogue channel. Occurs where voice-grade channels are multiplexed by the addition of a different carrier frequency to each channel. If a channel's carrier frequency is too high or too low, the signal will affect the next-higher or next-lower adjacent channel. The use of pulse code modulation (PCM) and T carriers eliminates crosstalk. See also *PCM* and *T carrier.*

CSPDN Circuit-switched public data network. See *circuit switching.*

CSU Channel Service Unit. Device furnished as an integral part of a digital access line where a user wishes to supply the bipolar signals. It provides the network with protection against user-side electrical anomalies (e.g., voltage surges) and provides the user with network clocking.

For a T1 facility, an important CSU function is to modify the outgoing data stream in such a way as to ensure a high enough density of 1 bits to ensure the proper operation of clock recovery circuitry. (For SDLC or HDLC transmission, this can also be accomplished by the re-inversion of an NRZI data stream, which will guarantee that at least every sixth bit is a 1 bit. See *NRZI*). See also *DSU*.

current loop Descriptive of a direct-current digital communication circuit on which all devices are in series, and in which 1s are represented by the flow of current and 0s are represented by the interruption of the current. Very rarely seen today, it was most commonly used for printing telegraphs (e.g., teletypewriters).

cursor A visual indication of the active position on a display terminal screen. Typically takes the form of an underscore or a reverse-video block and may either blink or remain steady (such attributes often being under user control).

customization The process by which a piece of software (e.g., a terminal emulator) is made to conform to the constraints of its operating environment and the requirements of its user.

Cyclic Redundancy Check See *CRC*.

D conditioning See *line conditioning*.

DACTLU See *SSCP-LU,* under *session*.

DACTPU See *SSCP-PU,* under *session*.

DARPA See *ARPANET*.

data access arrangement (DAA) Device once required by FCC regulations to be placed between a modem and the telephone circuit in order to protect the circuit against anomalies which might damage telephone company equipment. Such protection is still required, but is now incorporated in all DCEs which, in the United States, must comply with FCC Part 68 rules. Other countries have similar rules.

data chaining In SDLC, the chaining together of scattered segments of data into a complete SDLC frame.

Data Circuit-terminating Equipment See *DCE*.

data compression A process which reduces the number of bits which have to be sent over a data link. Compression may involve the elimination of repeated characters, the use of variable-length characters, etc.

data concentration The use of a single link for the transmission of data arriving over two or more links.

data entry 1. The act of entering data at a terminal.

2. Terminal usage in which the operation is characterized by repetition. Typically, data from paper documents is entered into fields on a formatted screen, then sent to the host through the depression of a key (e.g., Enter) which initiates transmission. Where transmission is very frequent, the use of the terminal is said to be transaction-intensive.

Data Flow Control See *SNA*.

data link The equipment configuration enabling two stations to communicate directly. It includes the paired DTEs, DCEs, and all intervening facilities.

Data Link Control See *SNA*.

Data Link layer See *OSI Model*.

data set 1. AT&T term for modem or DSU.

2. Term used in IBM's MVS environment (which includes TSO and ISPF) to describe a file or a collection of files (partitioned data set).

data stream Data that is flowing from one point to another in a network. A data stream usually consists of a succession of messages (or data blocks).

data streaming A protocol for transmitting data at high speed over a host system channel, in which the sending device maintains the channel in a transmit state for an extended period of time. The IBM 3174 Establishment Controller uses data streaming to transfer data at 2.5 Mbyte/s. The alternative (normal) transfer mode is called *interlocked* mode. In interlocked mode, the IBM 3174 transfers data at 1.25 Mbyte/s.

datagram 1. For the Internet protocol (IP) of TCP/IP, the basic unit of information passed across the Internet. Provides the basis for connectionless, best-effort packet delivery service. Also called *IP datagram*.

2. In a packet-switched network, a message whose size does not exceed the capacity of a single packet.

DCA See *DIA/DCA*.

DCE Data Circuit-terminating Equipment. Equipment (e.g., modem), used to connect Data Terminal Equipment (DTE) to a communications facility (e.g., telephone circuit). Sometimes also known as Data Communications Equipment (ambiguous term, now obsolete).

DDCMP Digital Data Communication Message Protocol. Byte count-oriented data link protocol defined by Digital Equipment Corporation (DEC). DDCMP messages have a formal structure in which a header contains class, count, response number, sequence number, and address fields. The count field specifies the number of octets (bytes) in an information field. A cyclic redundancy check (CRC) is performed separately on the header and on the information field.

DDS 1. Dataphone Digital Service. Digital network, owned by AT&T, offering switched digital communication links at 56 kbit/s, using a modification of the techniques defined in the now-obsolete CCITT Recommendation V.35.

2. See *DIA/DCA*.

dedicated line Leased or private communication line. Contrast with *dial-up line*.

device buffer IBM term for the area of memory within a terminal in which is stored the information that is to be displayed or printed. Thus, a device buffer may be a screen buffer or a printer buffer.

DIA/DCA Document Interchange Architecture/Document Content Architecture. IBM software and standards, designed to facilitate the use of documents in a variety of environments (host, PC, etc.) and their transfer among those environments. DIA consists of software, which provides the four following services:

- DDS—Document Distribution Services
- DLS—Document Library Services
- FTS—File Transfer Service
- APS—Application Processing Services

DCA defines a data stream in which the three following standard forms are used for the transfer of text:

- RFT—Revisable Form Text
- FFT—Final Form Text
- MFT—Mixed Form Text

dial-up line Communication line accessible via dial-up facilities, typically the public telephone network. Also called *switched line.*

digital loopback See *loopback test*.

DISOSS Distributed Office Support System. IBM application, running under CICS/VS, providing message handling, message switching, and document and file transfer services.

distortion An unwanted modification to a signal, caused by one or more of a number of factors. The most common types of distortion in data communication are:

attenuation On voice-grade lines, the amplitude of a signal degrades with distance. The degradation is greater at the higher and lower frequencies (toward 300 Hz and toward 3300 Hz) than in mid-range.

compandor Caused by the mismatching of a pair of compandors. See separate *compandor* entry.

envelope delay Caused by variations in the propagation times of different frequencies. Different frequency components of a signal will arrive at different times, although transmitted at the same time.

frequency shift Caused by minor differences between the modulation and demodulation radio-frequency carriers over microwave links.

harmonic (intermodulation *or* cross-modulation) Caused by non-linearities in the channel, specifically the clipping or limiting of the signal. Second and third harmonics (2 and 4 times the original frequency) of the fundamental signal are the most common.

jitter (*or* phase jitter) Caused by power line harmonics and perceived in the form of minor phase changes. Can be measured, using a 1004 Hz test tone and observing, on an oscilloscope, the forward and backward movement of the zero crossings of the received sine wave.

mark/space May be caused by cable capacitance at the DTE-DCE interface or by modem anomalies. In the case of high cable capacitance, signal rise and fall times are prolonged. In the case of modem anomalies, a negative bias will cause mark (1) bits to be longer than space (0) bits, and a positive bias will cause the opposite condition. Whatever the cause, the receiver circuitry, which normally samples the signal in mid-bit, may either fail to detect a polarity change or may detect it at the wrong time.

See also *noise*.

DLE Data Link Escape. Special character used in binary synchronous communications (BSC), in transparent mode, to signify that the following character is to be treated as a control character rather than as data, except if the following character is also DLE, in which case it is treated as a data character.

DLS See *DIA/DCA*.

DNA Digital Network Architecture. Network architecture defined by Digital Equipment Corporation and implemented on their equipment.

Since its introduction, in 1976, DNA has passed through four phases and is now at Phase V. DNA Phase V is compatible with both OSI and with DNA Phase IV. This is illustrated in the following DNA/OSI layer diagram, which has the OSI layer numbers on the right.

DNA/OSI Layers

OSI Application
DNA Application
Application 7
Presentation 6
Naming Service
DNA Session Control
Session 5
Transport 4
Network 3
Data Link 2
Physical 1

The DNA Session Control layer supports existing DNA applications and the use of Naming Services, which maps network resources and their addresses to plain-language names.

DOS/VSE See *VSE*.

double-current Using both positive and negative voltages, with no significance associated with a zero-voltage condition. DTE-to-DCE interface circuits are usually double-current.

downloading The transfer of information from one device to another, usually subordinate, device. Normally used to describe the transfer of a file or of executable software from a host system to a workstation or, for example, a cluster controller. Also used to describe the process of loading microcode to a microprocessor-based interface board.

DPM See *DPSK*.

DPSK Differential phase-shift keying. Modulation technique in which the value of a bit group is represented by a phase change in a sinusoidal wave. The phase change is relative to the phase angle which existed at the time of recognition of the previous bit group. Bit groups are typically dibits (two successive bits) or tribits (three successive bits). Also known as differential phase modulation (DPM). For typical phase changes, see *201 series* and *208 series* under *Bell modems*.

DSE Data switching exchange. A node in a circuit-switched or packet-switched network. The following illustration shows its relationship to DTEs and DCEs. The DSE enclosed in the broken-line box represents zero or more intermediate DSEs. CSUs, where used, would be between the DCE and DSE at either end.

DTE/DCE/DSE Relationships

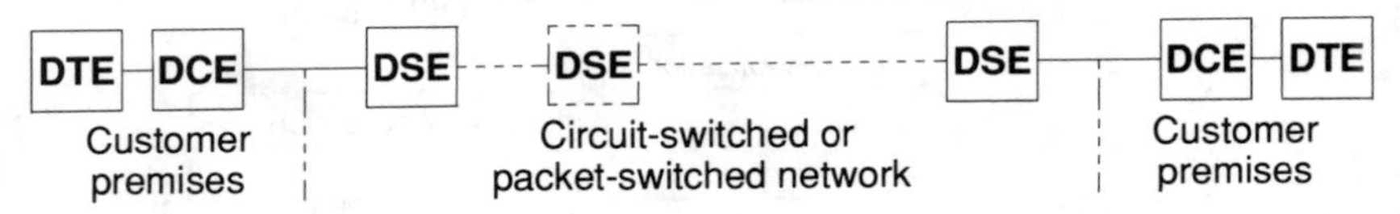

DSU 1. Data Service Unit. DCE used with digital communications circuits. Converts signals between those used at the DTE's serial interface (e.g., EIA-232-D, CCITT V.10, V.11, V.28, etc.) and the bipolar signals used on the digital network. Compare with *modem,* which is used with voice-grade circuits.

2. Distribution Service Unit. The element in SNA/DS with which an application transaction program communicates. See *SNA/DS*.

DTE Data Terminal Equipment. Device at which data transmission originates or terminates. May be a keyboard/display terminal, a printer, a computer, a communication controller, or any similar device.

DTMF Dual-Tone Multi-Frequency. An analogue signalling technique in which pairs of tones are simultaneously transmitted. Tone pairs consist of a row tone and a column tone. DTMF is used in Touch-Tone telephones, which have four row tones (697, 770,

852, and 941 Hz) and three column tones (1209, 1336, and 1477 Hz). A fourth column tone (1633 Hz) is used on military telephones. All eight tone frequencies are specified in CCITT Recommendation Q.23.

In addition to its use as a replacement for pulse dialling, DTMF signalling is used for a variety of consumer services, such as account balance inquiries (for bank accounts, credit cards, airline frequent-flier programs, etc.), for on-line credit authorization (in conjunction with magnetic-stripe readers attached to or associated with cash registers or other point-of-sale devices), and so on.

The following illustration shows a typical DTMF dial pad, with the optional military-use keys on the right. The letters associated with the numerals 2 to 9 reflect North American practice. CCITT Recommendation V.19 (see *Sec. 2*) deals with modems using the eight DTMF frequencies. CCITT Recommendation V.20 (see *Sec. 2*) defines an extended system using 12 frequencies.

DTMF Dial Pad

Frequency (Hz)	1209	1336	1477	1633 (Military use)
697	1	*ABC* 2	*DEF* 3	a
770	*GHI* 4	*JKL* 5	*MNO* 6	b
852	*PRS* 7	*TUV* 8	*WXY* 9	c
941	*	*OPER* 0	#	d

dumb terminal Popular term for a display terminal which has no local processing capability other than that associated with screen formatting.

duplex See *full-duplex*.

earth In electrical practice, a synonym of *ground*. Used in the United Kingdom and a number of other English-speaking countries, but not in the United States or Canada. Also earthed (grounded), earthing (grounding), and earth potential (ground potential).

EBCDIC Extended Binary Coded Decimal Interchange Code. Eight-bit code, defined by IBM, in 1964, with the introduction of the IBM System/360. EBCDIC includes values (code points) for control functions and for graphics (alphabetic, numeric, and special characters). Not all 256 possible values are assigned, and assignments are not always consistent (there being national variations, for example). The alphabetic and numeric subset corresponding to U.S. English practice is, however, consistent within all national usages based on the Roman alphabet. There are also non-Roman variations (e.g., Arabic, Cyrillic, Greek, Hebrew, Katakana, etc.). See *Sec. 5* for the full EBCDIC table and an explanation of the control codes.

ECF Enhanced Connectivity Facility. A set of IBM micro-to-mainframe programs used for file transfer, printer-sharing, virtual disks, and virtual file services. Under ECF, the PC-resident programs are known as *requesters* and the complementary host programs are known as *servers*. At present, ECF uses the Server-Requester Programming Interface (SRPI), via LU Type 2.0. In the future, ECF will use LU Type 6.2.

echo cancellation Modem technique allowing full-duplex operation over a two-wire circuit. After disabling echo suppressors, each modem goes through a short training period (in which it transmits a test pattern and receives the transmission's echo), allowing it to establish the delay and attenuation of the echo. Subsequently, at each modem, the transmitted signal will also be sent to an echo-cancellation stage of the receiver/decoder circuitry, which, after the known delay, will subtract it from the incoming signal. In this way, both modems can transmit simultaneously while each decodes only the other modem's signal.

echo suppressor Device used in voice communication links to eliminate the delayed (and, hence, distracting) "echoes" which can occur as a result of long-distance propagation delay. For full-duplex data communication, echo suppressors must be disabled (turned off). Full-duplex modems achieve this by initially transmitting a tone (typically, but not universally, 2100 Hz) for at least 400 ms. The 2100-Hz tone is specified by CCITT Recommendation V.25. To accommodate older standards, a modem may also transmit tones of 2225 and 2250 Hz. North American practice allows the tone to be anywhere in the range of 2010 to 2240 Hz (which accommodates both 2100 and 2225 Hz).

EHLLAPI Emulation High-Level Language Application Program Interface. IBM presentation-space (LU Type 2) application programming interface designed for use with high-level languages (C, COBOL, Pascal, BASIC, etc.). See also *LIM*.

EIA (Electronic Industries Association) standards A set of standards which includes, among others, data communications interface standards. The best known and most commonly implemented data communications standard is EIA-232-D, which defines circuit assignments and digital signalling levels for the DTE-to-DCE interface for serial-by-bit communication in start-stop and synchronous modes. A summary of all EIA standards relating to data communications is provided in *Sec. 2*.

End Chain See *chaining protocol*.

end node See *APPN*.

ending flag field See *flag field*.

Entry Assist A customizable feature of IBM 3174, 3274 and equivalent cluster controllers which, in conjunction with a CUT-mode terminal, provides local control of margins, word wrap, end-of-line warnings, and tab stops.

Entry Point See *NetView*.

equalization Process, within a modem, by which it compensates for certain predictable kinds of signal distortion.

adaptive equalization Modem capability providing for dynamic adjustment of attenuation equalization, delay equalization, or both. Older modems without this capability had to be set up according to the expected length of the data link. This was practical for leased lines, but not for dial-up connections. Sometimes called automatic adaptive equalization.

attenuation equalization Selective amplification of lower (toward 300 Hz) and higher (toward 3000 Hz) received frequencies to compensate for nonlinear signal degradation (attenuation distortion). Some modems use selective preemphasis on the transmitted signal to compensate for the expected nonlinear degradation.

delay equalization Selective delaying of various frequency components of the received signal to match the delay of other frequency components caused by envelope delay distortion. See also *distortion*.

equalizer Device, within a DCE, which performs the equalization function. See *equalization*.

error-detecting code A code so structured that transmission errors will cause detectable violations in the structure. See *constant-ratio code*.

escape ASCII control character used, usually within terminals and printers, to denote the start of a control function.

escape sequence An escape character (usually) and following characters which cause a device to perform a control function [e.g., select new character set (printer), clear screen (display), etc.].

ESCON Architecture Facility Enterprise Systems Connection Architecture Facility. IBM host-based facility associated with the ES/9000 (System/390) family. It consists of the following three elements:

- Protocols for high-speed, long-distance information exchange
- High-speed switched point-to-point communication
- Fiber-optic links (ESCON channels) operating at up to 17 Mbyte/s, for processor-to-processor or processor-to-peripheral-device communication at distances up to 60 km

ESCON channel See *ESCON Architecture Facility.*

ETX End of Text. BSC control character, signifying the end of a string of text characters.

explicit route SNA term for the portion of a path which lies between an originating and a receiving subarea node.

facility The kind of connection (e.g., public telephone network) which allows two stations to communicate with one another.

FCC Federal Communications Commission. U.S. Government agency, responsible for regulating most forms of electrical and radio communications within the United States.

FCS Frame Check Sequence. SDLC and HDLC term corresponding to CRC. The FCS polynomial function differs from that of CRC, but the purpose is the same.

FDDI Fiber Distributed Data Interface. Interface based on the use of fiber-optic connections, operating at a speed of 100 Mbit/s.

FDX See *full-duplex.*

FEP Front-end processor. See *communication controller.*

FEC Forward error correction. Technique in which a message contains enough redundant information to not only allow errors to be detected but also to correct a significant number of errors. FEC is not normally used in DTE-to-DTE data communications. However, it is used in a number of high-speed modems (DCEs). See also *ARQ.*

FFT See *DIA/DCA.*

figures shift 1. Code point, in teletypewriter practice, indicating that subsequent shift-sensitive codes represent numerals and certain other nonalphabetic characters. Baudot code and International Telegraph Alphabet No.2 code, both of which are five-level codes, include a figures-shift code point.

2. The mode in which a teletypewriter prints numerals and certain special characters.

See also *letters shift.*

First in Chain See *chaining protocol.*

flag field In HDLC and SDLC practice, an 8-bit value with the unique bit configuration 01111110. It is used to signify both the start (beginning flag field) and the end (ending flag field) of a frame. See also *HDLC frame, SDLC frame, synchronous,* and *zero-bit insertion.*

Focal Point See *NetView.*

four-of-eight code The best-known constant-ratio code, which uses 8 bits, of which 4 are always 1s and 4 are always 0s. This code was implemented in the 1960s, on IBM transmission control units (communication controllers), for communication between systems with a 6-bit character structure (64 possible values). There are 72 possible permutations of 4 bits out of 8, providing for 64 data values and eight control codes.

frame HDLC and SDLC term for a message block. See also *HDLC frame* and *SDLC frame.*

frame ground 1. Interface circuit connection between the chassis of a DCE and the chassis of a DTE. Required for reasons of safety. See also *CCITT Recommendation V.24* in *Sec. 2.*

2. Any grounding connection between or among the chassis of

two or more devices.

frame relay Packet-switched service option in which information frames are relayed to their destination with no error resolution at intermediate nodes. Takes advantage of the very low probability of transmission errors in current digital communication facilities. Where transmission errors occur, retransmission must be end-to-end.

frequency-shift keying (FSK) Modulation technique, used by low-speed anisochronous (asynchronous) modems, in which a tone of one frequency is used to represent the value 0 and a tone of another frequency is used to represent the value 1. At rates up to 300 bit/s, the originating modem uses one tone pair and the answering modem uses another tone pair, allowing full-duplex communication over a two-wire dial-up line. (Each modem can simultaneously modulate its transmitted data to its tone pair and demodulate received data that has been modulated to the other modem's tone pair.)

FTP See *TCP/IP*.

FTS See *DIA/DCA*.

full-duplex A two-way transmission scheme in which information can be carried in both directions simultaneously.

function base IBM term referring to the basic capability of a machine (without optional features). Often used in comparing the basic features of a new machine with those of a predecessor machine.

Function Management See *SNA*.

gateway Network node which operates as an interface between different network types (e.g., between a local-area network and an SNA network).

gateway server See *gateway*.

Gbit/s Gigabits (10^9 bits) per second. Unit of measure for extremely high speed data communication.

Gbyte Gigabyte. For mainframe memory, 1,073,741,824 (2^{30}) bytes. For external storage (e.g., mainframe-attached direct-access storage, magnetic tape, etc.), 1 billion (10^9) bytes.

GOSIP Government OSI Profile. Network management specifications, developed by the United States and United Kingdom governments and based on emerging ISO standards. See also *ISO* and *OSI*.

ground See *frame ground* and *signal ground*.

guard band Excess bandwidth used to minimize the possibility of crosstalk between adjacent channels. For carrier-based voice-grade channels, the excess bandwidth is created by the use of carrier-frequencies that are separated by more than the bandwidth of each channel. See also *crosstalk*.

half-duplex (HDX) A two-way transmission scheme in which information can be carried in only one direction at a time.

half session In SNA, this refers to the user-oriented functions at one end of an interaction. A pair of half sessions provides end-to-end services over a communications link and forms a session.

handshaking All the functions which must be completed before two machines or processes can communicate with each other.

harmonic distortion See *distortion*.

Hayes-compatible modems Modems incorporating the characteristics of the Smartmodem series manufactured by Hayes Microcomputer Products, Inc., especially the AT Command set used for modem control, testing, and automatic dialling. The best-known of the Smartmodem series is the Smartmodem 2400, which can operate as a Bell 103 or 212 series or in conformance with CCITT Recommendation V.22 or V.22 *bis*. The slower Smartmodem 1200 lacks only the V.22 *bis* capability. More recent Smartmodem 2400 models incorporate MNP, V.42 and V.42 *bis* capabilities, plus LAPB support for X.25 packet network connections. Although not compatible with CCITT Recommendation V.25 *bis*, the Hayes AT Command set has become a de facto industry standard. See also *MNP* and, in *Sec. 2, CCITT Recommendations V.22, V.22 bis, V.25 bis, V.42 and V.42 bis*.

HDLC High-level Data Link Control. ISO term for the link access protocols (LAPs) defined in ISO 3309:1984 and ISO 4335:1987, and in CCITT Recommendation X.25 and the ANSI X3.66 standard.

HDLC frame A transmission frame consisting of beginning and ending flag characters, an address field, a control field, and optionally, an information field.

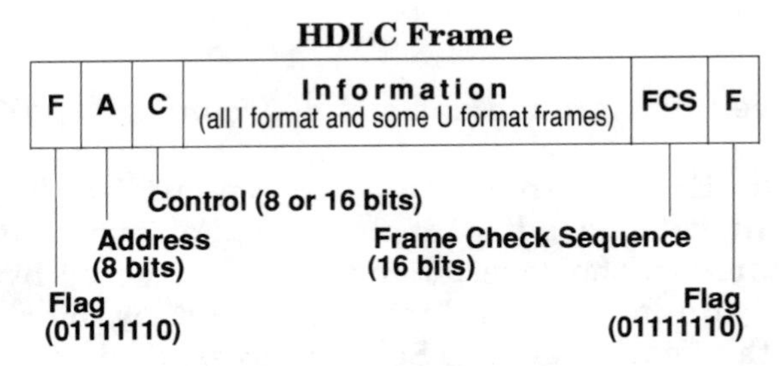

Each flag is a unique sequence of 8 bits (01111110) which cannot occur anywhere else in the frame (see *zero-bit insertion*). Flags allow receiving devices to recognize the beginning and end of a frame.

The frame check sequence is used for error detection. It is generated by the transmitting device and regenerated, for checking purposes, by the receiving device. See *FCS*.

The address field identifies the intended receiver of a command frame and the transmitter of a response frame. Four address values are defined—two for single link operation (LAP or LAPB), two for multilink operation (LAPB only). Multilink frames contain an additional 16-bit multilink control (MLC) field at the beginning of the information field. The four address values define command and response frames in the following manner:

11000000 Single link operation, DCE to DTE commands and DTE to DCE responses.

10000000 Single link operation, DTE to DCE commands and DCE to DTE responses.

11110000 Multilink operation, DCE to DTE commands and DTE to DCE responses.

11100000 Multilink operation, DTE to DCE commands and DCE to DTE responses.

(The address field of an *SDLC frame* is used in a completely different manner.)

The information field may contain end-user data, link access protocol information or, in the case of a multilink frame, both.

There are three frame formats, recognized by the leading bit or leading 2 bits of the control field.

I (information transfer) format Used to perform information transfer (end-user or protocol-dependent).

I Format Control Field (8-bit)

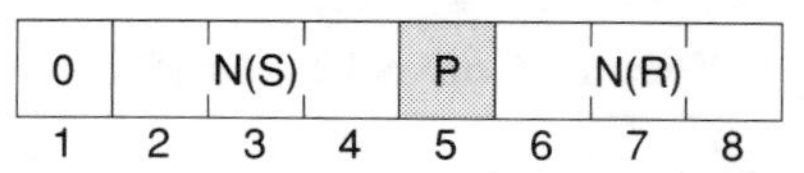

I Format Control Field (16-bit)

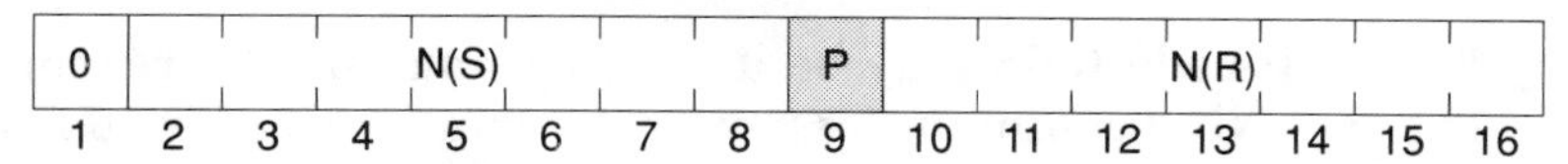

The send sequence number [N(S)] is incremented, modulo 8 (modulo 128 for 16-bit control fields), for each I format frame transmitted. The receive sequence number [N(R)] indicates the number of the next I or S format frame the station expects to receive (i.e., it is 1 higher than the last one received).

S (supervisory) format Used for supervisory commands and responses, identified by a 2-bit code. It does not contain an information field.

S Format Control Field (8-bit)

1	0	Code		P/F	N(R)		
1	2	3	4	5	6	7	8

S Format Control Field (16-bit)

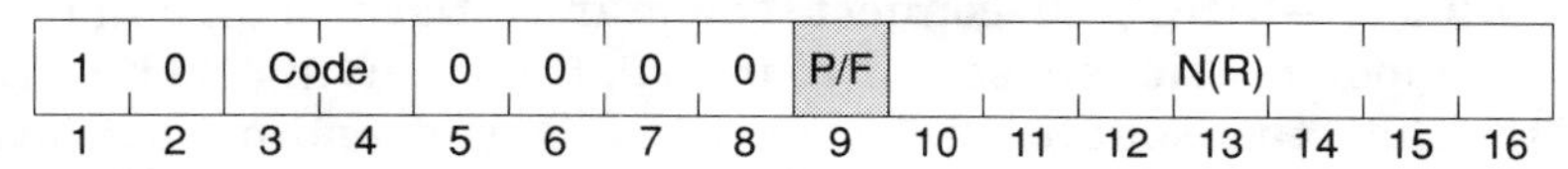

There are three possible codes.

00—RR (receive ready) is used by a DTE or DCE to indicate that it is ready to receive an I format frame. The N(R) value confirms the last received frame sequence number (i.e., that frames up to and including N(R)–1 have been received).

10—RNR (receive not ready) is used by a DTE or DCE to indicate a busy condition.

01—REJ (reject) is used by a receiving station to indicate a frame sequence error to the sending station. The N(R) value indicates the frame number at which retransmission is to begin and implicitly confirms satisfactory receipt of frame N(R)–1.

U (unnumbered) format Used for data link management, including the reporting of procedural errors that are not recoverable by retransmission. The information field may be used for data link management data. Commands and responses are identified by a 5-bit code.

U Format Control Field (8-bit only)

1	1	Code		P/F	Code (cont'd)		
1	2	3	4	5	6	7	8

The 5-bit code field may be used for commands or responses. The following command/response descriptions apply to **LAPB** (Link Access Protocol—Balanced) operation.

11 100—SABM Set asynchronous balanced mode (command). Places the addressed DTE or DCE in an asynchronous balanced mode (ABM) in which all control fields will be 8 bits in length. There is no information field.

11 110—SABME Set asynchronous balanced mode extended (command). Places the addressed DTE or DCE in an asynchronous balanced mode in which I format and S format control fields will be 16 bits in length. There is no information field.

Note that, for public data networks, mode of operation [modulo 8 (basic mode) or modulo 128 (extended mode)] is a subscription-time option.

00 010—DISC Disconnect (command). Used to terminate the mode previously set, or to inform the receiving DTE or DCE that the sending DTE or DCE is suspending operation. There is no information field.

00 110—UA Unnumbered acknowledgment (response). Used as an affirmative response to mode setting commands. A DTE or DCE receiving a mode setting command takes no action until it has sent the UA response. There is no information field.

10 001—FRMR Frame reject (response). Sent by a DTE or DCE to report an error condition. Includes a 24-bit (modulo-8 operation) or 40-bit (modulo-128 operation) information field.

11 000—DM Disconnect mode (response). Sent by a DTE or DCE to indicate that it is logically disconnected from the data link. There is no information field.

The following command/response descriptions apply to **LAP** (Link Access Protocol) operation. LAP supports 8-bit control fields only.

11 000—SARM Set asynchronous response mode (command). Places the addressed DTE or DCE in an asynchronous response mode (ARM) in which all control fields will be 8 bits in length. There is no information field.

00 010—DISC Disconnect (command). (See DISC description for LAPB, above.)

00 110—UA Unnumbered acknowledgment (response). (See UA description for LAPB, above.)

10 001—CMDR Command reject (response). Sent by a DTE or DCE to report an error condition not recoverable by means of a frame retransmission. Includes a 24-bit information field.

Poll/Final (P/F) bit The use of the Poll/Final bit for both LAP and LAPB operation is described, in detail, in CCITT Recommendation X.25 (*CCITT Blue Book, Volume VIII, Fascicle VIII.2,* ISBN 92-61-03671-6).

HDX See *half-duplex*.

hexadecimal Notation used to express the 16 possible values of a 4-bit group. IBM introduced hexadecimal notation in 1964, with the announcement of System/360. Pairs of hexadecimal digits are often used to express the value of a byte (8 bits).

Hexadecimal/Binary Conversion

0	0000	4	0100	8	1000	C	1100
1	0001	5	0101	9	1001	D	1101
2	0010	6	0110	A	1010	E	1110
3	0011	7	0111	B	1011	F	1111

high-speed circuit switching Circuit switching in which calls are established in considerably less than 1 s. Desirable for applications which make a separate call for each transaction.

HLLAPI High-Level Language Application Program Interface. Application programming interface designed for use with high-level languages (e.g., COBOL, Pascal, BASIC, etc.). See also *EHLLAPI* and *LIM*.

host-addressable printer A printer, operating as an NAU (Network-Addressable Unit) and capable, therefore, of accepting a data stream from a host computer.

host computer The primary computer in a network or a network domain. In SNA, the computer or one of the computers in which the SSCP resides.

hot key A key or key combination, on a PC or terminal keyboard, used to invoke immediate switching from one session or state to another session or state. Key combinations used for hot-keying are chosen to avoid conflict with keystrokes used to invoke terminal or PC-application functions. Hot-keying is frequently used in conjunction with TSR (terminate and stay resident) terminal emulation programs.

hub polling Polling protocol in which polling messages are sent from station to station and responses are sent directly to the primary station. Used extensively in airline reservation networks.

Huffman encoding A data compression technique in which shorter sequences of bits are assigned to the more frequently occurring code values, and longer sequences of bits are assigned to the less frequently occurring code values.

IBM See *Sec. 3* and *4*.

ICMP See *TCP/IP*.

IDSE International data switching exchange. See *DSE*.

IEEE Institute of Electrical and Electronic Engineers. IEEE has established a number of standards related to data communications, especially in the area of interfaces and local-area network protocols. For a summary of a number of IEEE standards, see *IEEE Data Communications Standards* in *Sec. 2*.

IMS, IMS/VS, IMS/ESA Information Management System (VS = Virtual Storage, ESA = Enterprise Systems Architecture). IBM mainframe-based database/data communications application subsystem.

information field The main, variable-length field in an HDLC or SDLC frame, containing either data or control information (associated with commands and responses). Not all HDLC and SDLC frames contain an information field. See also *BTU*.

interactive mode Mode of operation between DTEs in which the dialogue consists of a series of alternating entries and responses. Sometimes called conversational mode. Contrast with *batch mode*.

interchange code A code intended as a standard for the interchange of data among similar or dissimilar devices. See *ASCII* and *EBCDIC*. See also *Sec. 5*.

Internet Usually called "The Internet," this is a cooperative collection of networks which includes ARPANET, MILNET (Military Network—separated from ARPANET in 1984) and NSFnet (National Science Foundation Network). The Internet uses the TCP/IP protocol suite (IP means Internet Protocol). See also *TCP/IP*.

inverse video See *reverse video*.

IPDS Intelligent Printer Data Stream. See *IPDS* in *Sec. 4*.

ISDN Integrated Services Digital Network. A digital communications network in which data, voice, facsimile, graphics, video, etc., may be carried over a common physical circuit. ISDN is also used to refer, generically, to the service offered by such a network and, specifically, to the standards defined for its operation.

CCITT Study Group XVIII (Digital Networks) has developed a number of Recommendations for the structure, operation, and maintenance of ISDNs and for internetwork interfaces. Their Recommendations fall into three major groups:

I.110-I.257 General structure and service capabilities.

I.310-I.470 Overall network aspects and functions, ISDN user-network interfaces.

I.500-I.605 Internetwork interfaces and maintenance principles.

The overall features and functions defined by the CCITT I-Series Recommendations are:

- End-to-end digital service*
- Interface structures for standardized access
- A Basic Rate Interface (BRI) for small users, providing two 64-kbit/s bearer-service channels and one 16-kbit/s "delta" channel (sometimes known as 2B+D†).

---

* For input to an ISDN circuit, signals from devices whose normal mode of operation is analogue (e.g., telephones) are converted to digital form through pulse code modulation (PCM). Receiving equipment must perform the complementary digital-to-analogue conversion.

† *Recommendation I.412* defines B-channels (bearer service channels) and D-channels (delta channels). B channels are intended for user information; the D

- A Primary Rate Interface (PRI) for large users, providing 24 (North America and Japan* or 31 (Europe) 64-kbit/s channels (23B+D or 30B+D)
- Clearly defined basic and supplementary services

Five of the recommendations of CCITT Study Group XI (Telephone Switching and Signalling) are also significant. They are:

- **Q.920** and **Q.921**, which define the OSI layer 2 (Data Link) protocol for the D channel (for both BRI and PRI).
- **Q.930**, **Q.931** (basic call control) and **Q.932** (supplementary services), which define the OSI layer 3 (Network) protocol for the D channel (for both BRI and PRI).

Switch systems with ISDN capabilities (in addition to circuit- and packet-switching functions) include the following:

- **5ESS**, from AT&T
- **AXE 10**, from Ericsson
- **DMS-100**, from Northern Telecom
- **E10-FIVE**, from CIT-Alcatel
- **EWSD**, from Siemens
- **GTD-5 EAX**, from GTE
- **GX5000 Global Switch**, from Mitel
- **INS**, from NTT (Nippon Telegraph and Telephone)
- **NEAX 61E**, from NEC America
- **SuperDigital**, from NTT
- **System 12**, from ITT

For further information on ISDN, see *CCITT I-Series Recommendations* in *Sec. 2*.

ISO International Organization for Standardization. One of the two primary international standards bodies (the other is CCITT) developing data communications and network standards. A list of all ISO standards relating to data communications starts on p. 146.

---

channel is intended for signalling. It also defines two classes of H-channel, neither of which is intended to carry signalling for circuit switching. H_0 channels operate at 384 kbit/s and H_1 channels operate at either 1536 kbit/s (H_{11}) or 1920 kbit/s (H_{12}). Higher-rate H-channels are also contemplated.

* For the PRI, the number of 64-kbit/s channels is determined by trunk group (T-carrier) capacity. T1 carriers, at 1.544 Mbit/s, are implemented in North America and Japan. 2.048 Mbit/s carriers (CEPT) are implemented in Europe.

isochronous Correct, but infrequently-used term for devices (e.g., modems) which operate under the control of a common clocking mechanism. This term allows a distinction to be made between the operational mode of the modems and the transmission mode of the data. Isochronous modems are informally referred to as synchronous modems. (Note that asynchronous (start-stop) data streams *may* be transmitted isochronously; synchronous data streams *must* be transmitted isochronously.)

ISPF/PDF Interactive System Productivity Facility/Program Development Facility. IBM mainframe software, operating in the MVS and VM environments, allowing users of IBM 3270 terminals or compatible equivalents to interact with TSO and to make use of a number of additional functions (including a comprehensive full-screen editor) via screen panels, rather than individual command lines. See also *TSO*.

ITU International Telecommunication Union. Specialized agency of the United Nations, functioning through seven organizational entities, including CCITT. See also *CCITT*.

JES Job Entry System. IBM host-based software system designed to handle input, scheduling, and output of work submitted, in batch form, at a remote terminal device. Specific implementations are JES2 and JES3. JES is IBM's successor to RJE (Remote Job Entry), which evolved, in turn, from HASP (Houston Automatic Spooling Program).

kbit/s Kilobits (1000 bits) per second. Unit of measure for high-speed data transmission (e.g., 19.2 kbit/s, 64 kbit/s).

kbyte Kilobyte. For internal computer storage (memory), 1024 (2^{10}) bytes. For external storage (e.g., disk, tape), 1000 bytes.

kbyte/s Kilobytes (1000 bytes) per second. Unit of measure commonly used for transfer rates to and from peripheral devices (e.g., 640 kbyte/s for a channel-attached IBM 3274 cluster controller).

Kermit File-transfer protocol, based on sequence-numbered, variable-length data packets and using an error detection (block check) and recovery scheme. Kermit was designed at Columbia University and is in the public domain.

keyboard driver A software module designed to process PC keystrokes as they occur and either cause immediate action to be taken (e.g., for hot keys) or place corresponding data in the DOS keystroke queue. Keyboard drivers are often able to recognize keystroke combinations (e.g., Ctrl-1) normally ignored by the DOS keyboard BIOS (basic input/output system) and enqueue unique values for them in the keystroke buffer.

keyboard/display session Host communications session in which information destined for the host is entered by means of keystrokes and information destined for the user is presented in a device buffer. Whether the keystrokes are real or simulated and whether the contents of the device buffer are actually displayed depends on whether the user is an operator or, via an application program interface (API), a program (which may, in turn, be communicating with an operator).

keyboard/display terminal Terminal capable of supporting one or more operator-to-host keyboard/display sessions. Some keyboard/display terminals (e.g., IBM InfoWindow 3472) can also support a host-addressable printer session using an attached printer.

LAP and LAPB Link Access Protocol and Link Access Protocol—Balanced. The data link layer functions of CCITT Recommendation X.25. They are implementations of HDLC.

Last in Chain See *chaining protocol.*

leased line See *dedicated line.*

leased-line conditioning See *line conditioning.*

letters shift 1. Code point, in teletypewriter practice, indicating that subsequent shift-sensitive codes represent alphabetic characters. Baudot code and International Telegraph Alphabet No.2 code, both of which are five-level codes, include a letters-shift code point.

2. The mode in which a teletypwriter prints alphabetic characters.

See also *figures shift.*

LIM Language Interface Module. Object program module designed to be linked with object modules produced by a high-level language compiler (e.g., C) to produce an executable application program for operation with a high-level language application program interface (e.g., EHLLAPI).

limited-distance modem DCE used for communication over a two-wire or four-wire private-line metallic circuit. Depending on the signalling method (typically NRZ bipolar or sinewave), may or may not require DC continuity. Typical distance limits range from 4 to over 35 km. Maximum data rate is inversely proportional to distance. Also known as a line driver.

line conditioning A process by which certain voice-grade leased (private) line quality attributes are guaranteed. U.S. telephone companies have traditionally offered up to five levels of line conditioning in two major categories—C and D.

C conditioning sets limits on white noise, impulse noise (spiking), frequency translation, envelope-delay distortion, frequency response and attenuation distortion. In ascending order of quality, the three levels of C conditioning are C1, C2, and C4.

D conditioning sets further limits on C-notched noise and second and third harmonic distortion and is required for the operation of 9600-bit/s modems such as the Bell 209 series. D1 conditioning applies to point-to-point operation; D2 conditioning applies to three-point (limited multipoint) operation.

Present-day modems do not normally require line conditioning. Through a combination of advanced modulation, data compression and forward error correction techniques, plus today's higher-quality telephone circuits, they are able to operate at effective rates up to 38,400 bit/s over leased or dial-up voice-grade lines.

See also *distortion* and *noise*.

line driver See *limited-distance modem*.

line turnaround For a half-duplex connection, the process of reversing the direction of transmission.

logical connection In SNA, a connection between two NAUs. Analogous to virtual circuit. See also *session*.

Logical Unit (LU) An "end user" in an SNA network. End users may be terminals (e.g., 3278), terminal emulation programs, or applications. LUs may communicate with the SSCP, with host applications (LU Type 0, 1, 2, 3, 4 and 7), or with other LUs of the same type (LU Type 6.0, 6.1 and 6.2 only). IBM has defined nine LU types.

Type	Description
0	User-definable Logical Unit. Used, in general, to support terminal types and characteristics not defined by any of the other numbered LU types. Requires specific user-written code at the host.
1	Host-addressable printer with SNA character string (SCS) support (e.g., IBM 3287).
2	3270-data-stream-compatible keyboard/display terminal (e.g., IBM 3278).
3	3270-data-stream-compatible printer, without SNA character string (SCS) support.

Type	Description
4	Peer-to-peer communication device with SNA character string (SCS) support.
6.0	Logical Unit for program-to-program communication between two CICS/VS applications on the same host.
6.1	Logical Unit for program-to-program communication between CICS/VS and/or IMS/VS applications on the same host or on different hosts.
6.2	Logical Unit for general-purpose program-to-program communication.
7	Host-addressable keyboard/display terminal with SNA character string (SCS) or 5250 data stream support.

loopback test A test in which a DCE (e.g., modem) receives its own transmitted signal, either locally or from a remote DCE. Loopback tests are used to identify a failing component in a data link. The information transmitted may be a test pattern generated within the DCE or may originate in the attached DTE (e.g., as a series of keystrokes). Where a test pattern is used, an error detector, within the DCE, compares the transmitted and received versions. Where the information originates in a DTE, the transmitted and received versions may be compared by diagnostic software or, in the case of a simple ASCII terminal (for example), may be visually checked.

local loopback As the following simplified illustrations show, the transmitted signal on the data circuit side of the DCE is fed back to the receiver. Where the DCE is a modem, the term used is *local analogue loopback.*

In a self-test operation, a comparator receives a test pattern (i.e., a fixed series of 1 and 0 bits) directly from a test-pattern generator and via the DCE's transmitter and receiver circuits. A mismatch indicates an internal failure, probably in the transmitter or the receiver circuitry.

Local Analogue Loopback Self-Test

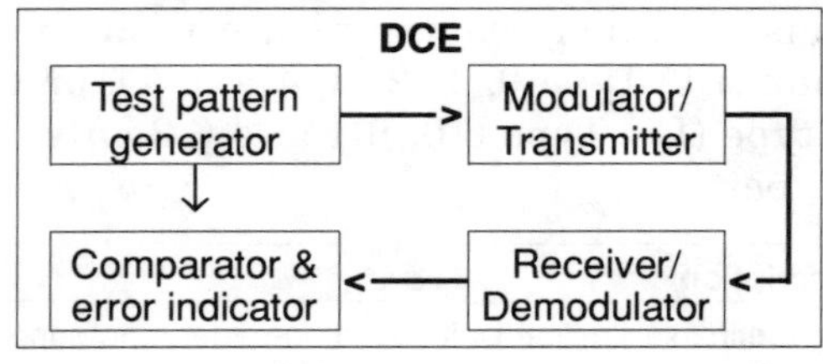

With a DTE-originated test, software in the DTE can serve as both test-pattern generator and comparator. In the case of a simple full-duplex terminal, the operator can enter characters whose correct display or printing will confirm a successful test. CCITT Recommendation V.54 refers to this as a Loop 3 test. (A Loop 1 test tests the DTE interface logic only.)

Local Analogue Loopback with DTE-Originated Test

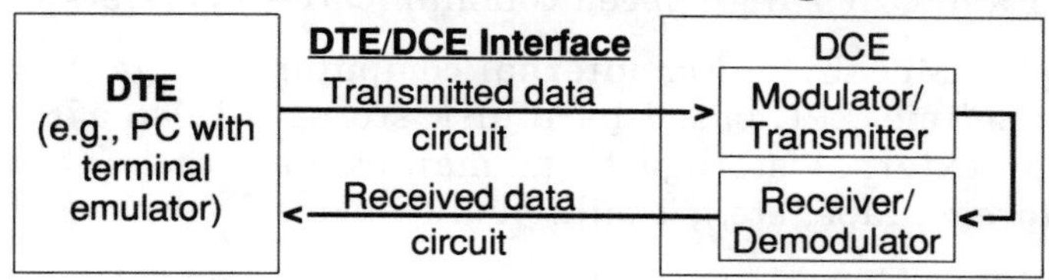

remote loopback Once a DCE has passed a local loopback test, a remote analogue loopback test can be run to check the communication circuit, followed by a remote digital loopback to check the remote DCE. In the following simplified illustration, the broken line within the remote DCE shows the path for the remote analogue loopback, and the solid line shows the digital loopback path. A DTE-originated test is shown. CCITT Recommendation V.54 refers to the analogue loopback test as a Loop 4 test and to the digital loopback test as a Loop 2 test. Remote loopback self-tests are very common.

Remote Analogue and Remote Digital Loopback

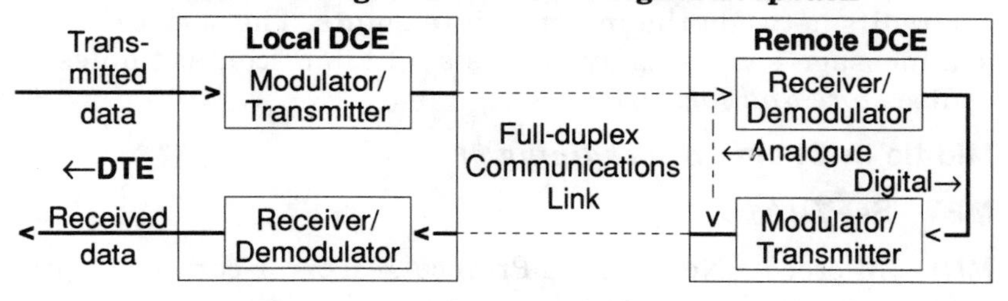

LRC Longitudinal Redundancy Check. A redundant (noninformational) character added to the end of a transmitted frame (block), in which each bit is 0 or 1, depending on whether the total number of 1 bits in the corresponding position of all preceding characters is odd or even, respectively (odd parity), or even or odd, respectively (even parity). Normally used with block-mode transmission of ASCII data (including ASCII-mode BSC).

LU See *Logical Unit*.

LU-LU session See *session*.

mark For an interchange circuit, a condition representing a value of 1. A mark condition lasting for *n* bit times represents *n* successive binary 1s. See also *space* and *start-stop*.

Mbit/s Megabits (10^6 bits) per second. Unit of measure for very-high-speed data transmission (e.g., 4 Mbit/s or 16 Mbit/s for Token Ring, 10 Mbit/s for Ethernet, 1.544 and 2.048 Mbit/s for T1 and CEPT high-speed communications carriers).

Mbyte Megabyte. For internal computer storage (memory) and, sometimes, PC-based hard disk storage, 1,048,576 (2^{20}) bytes. For external storage (e.g., mainframe-attached direct-access storage, tape, etc.), 1 million (10^6) bytes.

Mbyte/s Megabytes (million bytes) per second. Unit of measure commonly used for transfer rates to and from peripheral devices (e.g., 1.25 or 2.5 Mbyte/s for a channel-attached IBM 3174 Establishment Controller).

message An ordered collection of data in a form that can be processed by the receiver. In the simplest case, message may be synonymous with block or frame. In practice, a message may occupy only part of a frame or may span two or more frames.

message switching Descriptive of a process in which messages are stored at one or more intermediate points between sender and receiver. No end-to-end path (connection) is ever established and the progress of messages is governed by the availability of circuits between the intermediate points. The service offered by a message switching system is said to be "connectionless." See also *store and forward*.

Middle in Chain See *chaining protocol*.

MFT See *DIA/DCA*.

MNP Microcom Networking Protocol. An error-correcting protocol for modems. MNP has nine classes, of which Classes 1 to 4 are in the public domain, and Classes 5 to 9 are proprietary (Class 5 is widely licensed). The nine classes are defined in the following table.

MNP Classes

Class	Description
1	Half-duplex, start-stop, byte-oriented.
2	Full-duplex, start-stop, byte-oriented.
3	Full-duplex, synchronous, bit-oriented.
4	Full-duplex, synchronous, bit-oriented, with adaptive packet assembly (shorter packets on noisy lines) and data phase optimization (reduction in the number of header bits).
5	Same as Class 4, plus two types of data compression (run-length and Huffman encoding).

Class	Description
6	Same as Class 5, plus universal link negotiation (alteration of modulation methods) and statistical duplexing (bandwidth allocation based on traffic density). Class 6 operates on a half-duplex line, but simulates full-duplex operation.
7	Same as Class 6, plus adaptive encoding.
8	Same as Class 7, plus V.29 fast train capability and emulation of full-duplex transmission.
9	Same as Class 8, plus conformance with CCITT recommendation V.32. Throughput is up to three times that of a standard V.32 modem.

mode A given condition of functioning. In data communications, there are three major mode categories.

information mode For a message on a communications link, either control mode or text mode. Control mode is used for such functions as terminal addressing and selection; text mode is used for the transfer of data.

operating mode Describes the directional manner of use of a communications link. Three possible operating modes are simplex (one direction only), half duplex (one direction at a time), and duplex, or full duplex, (both directions at the same time).

synchronization mode Describes the level at which data on a communications link is synchronized. Start-stop mode denotes character-by-character synchronization; synchronous mode denotes frame-by-frame (or block level) synchronization.

modem Modulator/demodulator. DCE used to connect a DTE (e.g., terminal) to a voice-grade communications circuit. Converts the DTE's digital signals to analogue form for transmission (modulation) and, conversely, converts the received analogue signals from the voice-grade circuit to digital form (demodulation). See also *Bell modems* and *Hayes-compatible modems*.

modem carrier Continuous frequency transmitted by a modem and capable of being modulated on the basis of the bits in the data stream (from the DTE) and according to the method defined for the modem. Also referred to as carrier or carrier frequency.

modem eliminator Device used for isochronous local connection between two DTEs, each of which requires DCE clocking. Appears as a pair of full-duplex isochronous DCEs on a dedicated line.

modem sharing unit A device allowing two or more DTEs to be connected to the same modem. Typically used in conjunction with a multipoint line.

modulo *n* Having a maximum value of $n-1$. Used in sequence numbering, where a fixed field length is used for the sequence number. For example, a 3-bit field provides for eight possible values (2^3), forcing a reset to 0 after the count reaches 7. The sequence number for which the field is used is described as modulo 8. SDLC and HDLC information frames use either modulo 8 or modulo 128 (7 bits) sequence-number fields. See *SDLC frame*.

(The full mathematical expression is "i modulo n," where i is any integer and the expression means "the remainder after dividing i by n." Thus, 3 modulo 8 = 3; 45 modulo 8 = 5; 1234 modulo 128 = 82.)

multi-session The support of multiple communication sessions within one terminal or terminal-emulation device.

multidrop line See *multipoint line*.

multiplexer A device which allows multiple independent data streams to share a data link. Usually used in pairs over a point-to-point data link, as shown in the following illustration.

Point-to-Point Multiplexer Configuration

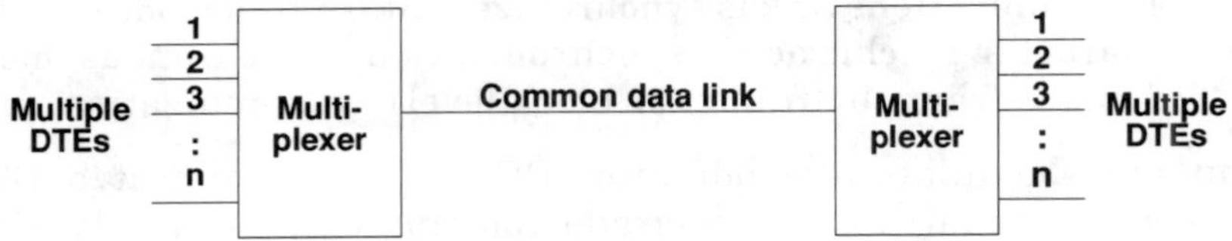

frequency-division (FDM) Uses frequency-shift keying (FSK), with a different mark/space frequency pair for each DTE-to-DTE channel. Can typically handle five 300-bit/s channels over a C1-conditioned leased line or six 300-bit/s channels over a C4-conditioned line. Because FSK is not code-sensitive, the interchange code does not have to be the same for all DTE-to-DTE channels. Many FDMs are modular and are assembled from as many "channel units" as required. FDMs can be used in point-to-point, point-to-multipoint, and multipoint-to-multipoint line configurations. Individual channels operate as point-to-point.

statistical (STM) Takes advantage of the intermittent nature of the transfer of data between a pair of DTEs, especially with interactive applications, allowing the support of a number of DTEs whose aggregate data rate exceeds the rate used over the data link.

An STM appears as a DCE to each of the attached DTEs, although it may or may not include the DCE function.

The STM identifies characters, character strings, or frames (for synchronous operation) coming from each DTE by appending a channel number header prior to their transmission over the data link. This requires buffering within the STM, with the obvious risk of exceeding the buffer's capacity when several terminals are transmitting at the same time. This is handled by a flow-control protocol whenever buffer occupancy passes a defined threshold. Flow control can be handled via the Clear-to-Send circuit or, for full-duplex operation, by means of flow control characters (start-stop) or frames (synchronous). For full-duplex start-stop operation and an ASCII data stream, the X-Off (DC3) and X-On (DC1) characters are used for flow control.

For incoming data, the channel number header is used to direct characters or frames to the appropriate DTE.

Like FDMs, STMs can be used over point-to-point or multipoint lines, although the latter requires the STMs to support a polling/addressing protocol (not necessary for FDMs). The following illustration shows an eight-channel primary STM supporting a four-channel and two two-channel secondary STMs on a multipoint line. To the DTE pairs, each channel appears as a point-to-point link.

The simpler STMs require that all attached terminals operate in the same mode (e.g., start-stop) and use the same interchange code (e.g., ASCII) and data link control protocol. More advanced STMs can be configured for multiple modes, interchange codes and protocols.

Statistical Multiplexers in Multipoint Configuration

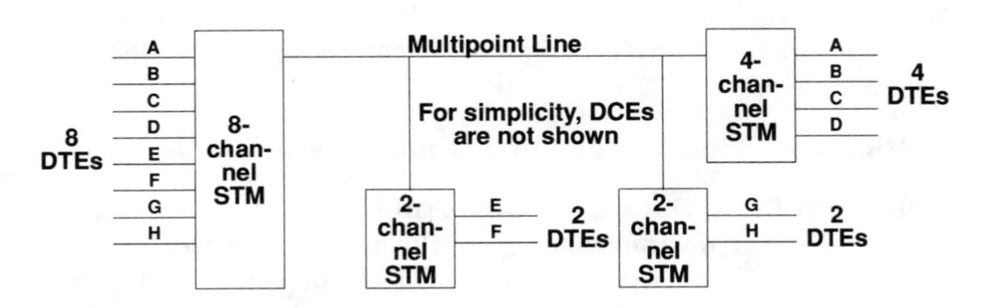

The term *concentrator* (or *line concentrator*) is often used instead of statistical multiplexer.

time-division (TDM) Operates isochronously and allocates a time slot equal to the transmitted bit time (over the link) to each of the supported channels in turn. The number of time slots allocated to a given channel depends on the bit rate of the pair of DTEs it serves. For example, a 9600-bit/s TDM (e.g., a Bell 209 or CCITT V.29 modem) supporting one 4800-bit/s channel and two 2400-bit/s channels would allocate two of every four time slots of 1⁄9600 s duration to the 4800-bit/s channel and one each to the 2400-bit/s channels. The following diagram illustrates this, showing the bits processed over five successive 1⁄2400-s periods . For simplicity, only one direction is shown. The arrows show the direction of flow and the starting point for numbering the bits is arbitrary.

Bit Flow between 9600-bit/s TDMs

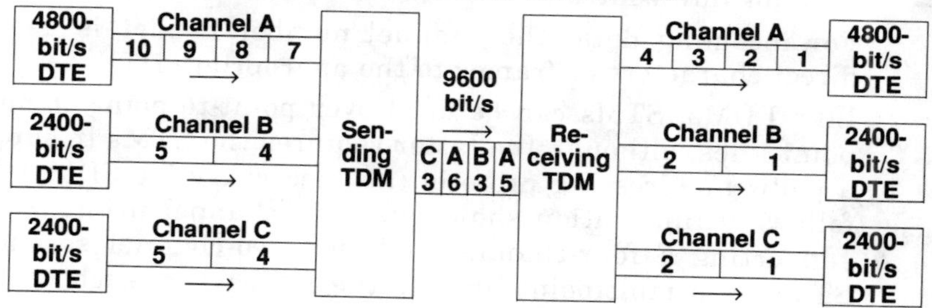

Some TDMs multiplex by the octet (byte), rather than by the bit. Those used for T1 and CEPT 2.048-Mbit/s carriers (see *T carrier*) combine a set number of channels (24 or 30), all operating at the same data rate (64 kbit/s). Such TDMs also operate at a bit rate greater than the sum of the bit rates of the constituent channels. This is because of overhead (framing bits).

multiplexing Technique whereby two or more data streams are concurrently transferred over a common channel. See *multiplexer*.

multiplexor Optional spelling of multiplexer.

multipoint line Data link supporting multiple DTE connections, typically but not necessarily with one DTE controlling the link by polling (querying) the other DTEs for input and addressing output to the other DTEs. With frequency-division or statistical multiplexers, a multipoint line can support multiple independent point-to-point channels. See also *point-to-point line*.

MVS Multiple Virtual Storage. IBM operating system for large host systems (typically IBM 4381, 308x, 309x, and System/390). Implementations of MVS include MVS/XA (Extended Architecture) and MVS/ESA (Enterprise Systems Architecture).

national use See *national use* in *Sec. 4*.

NAU Network Addressable Unit. A location in an SNA network which supports one or more ports for communication via the network. Each NAU has a network address. There are three NAU types, specifically SSCPs (System Service Control Points), PUs (Physical Units), and LUs (Logical Units).

NCP See *ACF/NCP*.

NetView IBM network management software for SNA networks. NetView runs as a VTAM application on an IBM System/370 or System/390 mainframe computer. Structurally, NetView is organized as a Focal Point (the VTAM-based component) and one or more Entry Points (typically IBM 3174 Establishment Controllers), Service Points (NetView/PC software running on PS/2s under OS/2 Extended Edition, or AIX NetView Service Point software running on IBM RS/6000 POWERservers or POWERstations, or SUN SPARCservers or SPARCstations) and, sometimes, Secondary Focal Points (typically on other mainframes or AS/400s).

NetView Structure (Simplified)

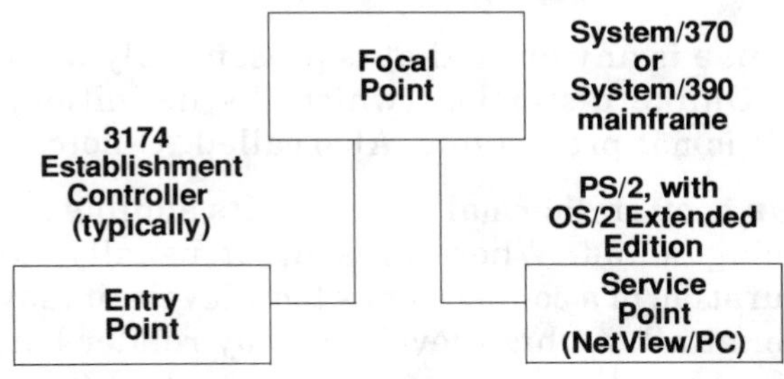

The Focal Point is the central point for collecting, analyzing, and storing network and system management data. It includes a Host Command Facility, Session Monitor, Hardware Monitor, Status Monitor, On-line Help Facility, Help Desk Facility, and a Browse Facility. It provides an application programming interface (API) for user-written extensions and NetView-supportive applications.

Entry Points support SNA applications and devices. They send alerts and response-time monitor information to the Focal Point and process commands and requests from the Focal Point.

Service Points are intended to support non-SNA devices or subnetworks (e.g., LANs). They require user- or vendor-written code to translate alerts and event-notification messages from non-IBM to IBM format. The incorporation, by most LAN vendors, of Entry Point software in their LAN-to-SNA gateways may render Service Points obsolete.

network A configuration in which two or more locations are connected for the purpose of exchanging data.

Network layer See *OSI*.

network node See *APPN*.

node A network element with communication links to two or more other network elements, which may be other nodes or end users.

noise Random variations in voltage, current, or data, resulting from a natural phenomenon, channel component deficiencies, or both. Except for notched noise, it is not dependent on the presence of a signal.

impulse noise is caused by random electrical interference, whose source may be an electrical storm, the operation of a switch, etc. If its occurrence is not too frequent, it causes only minor degradation in service (i.e., occasional retransmissions).

notched noise is any noise that is present only when a signal is present. Unlike distortion (which, by definition, requires a signal), it is not predictable. Also called "C-notched noise."

white noise is often thermal in origin. Its audible manifestation is a hissing sound. Where present, it usually exists for the entire duration of a connection. At low levels, it may not impair operation at all. At high levels, it may render both voice and data communication impossible. Also called Gaussian noise. Electrical noise, specifically of thermal origin, is also known as Johnson noise.

See also *distortion*.

non-SNA Non-Systems Network Architecture. Descriptive of any terminal or communications implementation which does not conform to SNA rules. One specific use of the term refers to the operation of remote BSC or byte-multiplexer channel-attached cluster or establishment controllers.

NRZ Non-Return-to-Zero. Encoding scheme, implemented at the DTE-to-DCE interface, in which a positive voltage is used to represent a binary 0 in the data stream, and a negative voltage is used to represent a binary 1 in the data stream.

NRZI Non-Return-to-Zero Inverted. Encoding scheme, implemented at the DTE-to-DCE interface, in which a polarity change (e.g., from positive to negative or from negative to positive voltage) occurs every time there is a binary 0 in the data stream. With SDLC and HDLC, zero-bit insertion ensures that there will never be more than six contiguous 1s; hence a polarity change will occur at least every 6 bits. This improves the reliability of bit synchronization, regardless of frame length. Also called zero-complemented differential encoding. (Contrast with *NRZ* encoding, with which a frame with a large number of successive 0 bits would not cause a polarity change until the occurrence of a 1 bit.) The use, in DCEs, of scrambler circuitry has reduced the necessity for NRZI encoding. See also *channel service unit* and *scrambler*.

null modem Connection allowing communication between a pair of start-stop DTEs. Uses connections which cause signals on the interchange circuits of each DTE to appear, to the other DTE, as DCE signals. A minimum null modem (for full-duplex operation only) establishes connections between the two DTEs' signal ground and frame ground circuits (connector pins 1 and 7) and connects the transmit circuit (pin 2) of each DTE to the receive circuit (pin 3) of the other DTE. The following illustration shows the connections required for full operation in either full- or half-duplex mode with switched (dial-up) circuit simulation.

A null modem may also be used to provide isochronous communication between a pair of synchronous DTEs. This depends on both DTEs providing Transmitter Signal Element Timing. Each DTE's timing circuit (pin 24) is connected to the other DTE's Receiver Signal Element Timing circuit (pin 17). If only one DTE provides a timing signal, it could, in theory, be used to drive the DTE's own Receiver Signal Element Timing circuit (pin 17), plus the Transmitter and Receiver Signal Element Timing Circuits (pins 15 and 17) of the other DTE. However, it is questionable whether enough power would be available to make such a setup reliable. See *modem eliminator*. See also *breakout box*.

Null Modem Connections

DTE A Connector

DTE B Connector

22 Ring (Calling) Indicator 22
20 Data Terminal Ready 20
8 Received Line Signal Detector 8
7 Signal Ground 7
6 Data Set Ready 6
5 Clear to Send 5
4 Request to Send 4
3 Received Data 3
2 Transmitted Data 2
1 Frame Ground 1

octet An addressable group of 8 bits. Term used by CCITT and other organizations. The convention for bit numbering within an octet is different from that used for a byte (which is an IBM term). See comments under *Notation,* in the *Preface.* See also *byte*.

Only in Chain See *chaining protocol*.

Open Document Architecture Consortium Group, composed of Groupe Bull, Digital Equipment Corporation, IBM, ICL, Siemens/Nixdorf Information Systems AG, and Unisys Corporation, formed in April 1991 to develop software for the electronic interchange of documents containing text, pictures, and diagrams. One of the group's goals is to promote ISO's Open Document Architecture standard.

OSI Model Layered communications architecture for Open Systems Interconnection. Seven layers are defined:

7	Application
6	Presentation
5	Session
4	Transport
3	Network
2	Data Link
1	Physical

Layers 2 to 6 are similar to the five layers of SNA, with the greatest similarity occurring in the Data Link layer.

The purpose of the OSI Model is to provide a framework for the development of mutually compatible products which involve communication between or among applications. Applications are defined very broadly and may include operator processes (e.g., the actions of users at terminals) in addition to program-based processes.

Standards for OSI implementation are published by the International Organization for Standardization (ISO) and are based on the CCITT X.200-Series Recommendations. *Sec. 2* of this book lists the CCITT Recommendations and provides a cross-reference to the corresponding ISO standards (also listed).

In spite of the historic dominance of IBM's Systems Network Architecture (SNA), the development and implementation of Open Systems is proceeding very rapidly, with IBM now playing a very significant role.

A summary description of each layer is provided in *Sec. 2,* under *CCITT Recommendation X.200.*

packet switching The routing and transmission of packets of data across a network, based on origin/destination address pairs contained within the packets. All packets containing the same address pair are said to be using the same "virtual circuit." See also *virtual circuit.*

PAD Packet Assembler/Disassembler. Device or program used to create packets of data for transmission over a CCITT X.25 packet data network and to remove data from the received packets. The most common PAD is the one defined by CCITT Recommendation X.28, used for packetizing and depacketizing start-stop data streams.

paper tape Continuous strip of paper of constant width into which holes are punched to record received data or to prepare recorded data for transmission. Depending on the application, four, five, six, or eight holes may be punched in lateral frames, each of which represents a character or a control code. For each frame, a hole represents a 1 and the absence of a hole represents a 0. Paper tape may be chadded (holes punched out completely) or chadless (chad for each hole still attached at leading edge). Chadded tape may be read optically or mechanically. Chadless tape, common in teletypewriter operation, is read mechanically.

Paper tape is still used in teletypewriters and related equipment. Older applications include computer input and output and the recording of cash register operations. Not all paper tape is punched. In some parts of the world, telegrams are still received on printed paper tape.

parallel transmission group Group of circuits, all of which are used to link the same pair of nodes. With appropriate software at each node, such a group of circuits offers the combined benefit of its aggregate bandwidth (and, hence, data rate) and redundancy. The failure of any circuit causes no more than a reduction in throughput.

parity check See *VRC*.

path In SNA, a series of nodes and communication links over which data must travel in order to get from one Logical Unit to another.

Path Control See *SNA*.

PCM Pulse Code Modulation. A means of converting analogue signals to digital form, allowing the use of digital communications facilities such as ISDN circuits and T carriers. The analogue signal is sampled 8000 times per second to determine its amplitude, which is converted to an 8-bit value representing one of 256 possible levels. The resulting data rate is 64,000 (8000 x 8) bit/s. See also *ADPCM*.

PDF Program Development Facility. See *ISPF/PDF*.

PDN Public Data Network. Network operated by a PTT (see definition) or other enterprise, providing switched connections either primarily or exclusively for the transmission of data. The corollary term for the traditional telephone system would be public voice network. See also *circuit switching* and *packet switching*.

PEL Picture element. IBM term for pixel, which is the smallest addressable element of a graphics display.

peripheral node Cluster Controller, Establishment Controller, or remote IBM 5250 controller (e.g., 5394). Can communicate only with the subarea node to which it is attached. See also *physical unit*.

permanent virtual circuit See *virtual circuit*.

physical address See *physical address* in *Sec. 4*.

Physical layer See *OSI*.

Physical Unit (PU) A node in an SNA network, supporting one or more Logical Units (LUs). There are five physical unit types.

Type	Description
1	Peripheral node. Can be a remote twinaxial control unit (IBM 5294 or 5394) or the obsolete IBM 3271 Model 11 or 12 SDLC cluster controller.
2.0	Peripheral node, specifically an IBM 3174 establishment controller or 3274 cluster controller (or equivalent). Not capable of establishing LU-LU sessions without the help of an SSCP. A PU Type 2.0 can communicate only with a PU Type 4. (Previously referred to as PU Type 2.)
2.1	Peripheral node, incorporating the characteristics of PU Type 2.0 and having the additional capability of supporting one or more LU Type 6.2s. A PU Type 2.1 can communicate with a PU Type 4 or another PU Type 2.1. The major difference between PU Type 2.0 and PU Type 2.1 is the addition of a "Peripheral Node Control Point" and a "Link Manager." PU Type 2.1 can reside in any programmable device.
4	Subarea node, specifically an IBM 3705, 3720, 3725, or 3745 communication controller or equivalent, running ACF/NCP. A PU Type 4 can communicate with all other PU types and with other PU Type 4s.
5	System/370- or System/390-class host processor, running ACF/VTAM and including an SSCP. Also, System/36, System/38, or AS/400, supporting 5250 or 3270 Information Display Systems.

PIU Path Information Unit. In SNA, a unit of information consisting of a BIU (Basic Information Unit) and a TH (Transmission Header) added by the Path Control layer.

.... Path Information Unit ....

TH	Basic Information Unit (BIU)

See also *BIU, BTU, RH, RU,* and *TH*.

PLU See *Primary Logical Unit*.

point-to-point line Data link supporting communication between only two DTEs. See *multipoint line*.

Poll/Final (P/F) bit See *HDLC frame* and *SDLC frame*.

polling Process of querying stations (DTEs) on a multipoint line, one at a time, for pending messages, operational status, etc. See also *automatic polling*.

polling list List of station addresses on a multipoint line, used in polling.

Polycenter Network management product line from Digital Equipment Corporation.

port 1. In SNA, a point of connection to an NAU (Network-Addressable Unit).

2. A point at which a device can be connected to a computer,

communication controller, cluster controller, or other equipment in a network for the purpose of sending or receiving data.

Presentation layer See *OSI Model.*

Presentation Services In SNA, the function, within a logical unit, which handles the display, printing, editing, mapping, or acceptance of data. Presentation services manage the presentation space. The SNA layer, previously referred to as Presentation Services, is now called Function Management. See also *SNA*.

presentation space See *presentation space* in *Sec. 4*.

PRI Primary Rate Interface. See *ISDN*.

Primary Logical Unit (PLU) Usually a host application which, in communication with a Secondary Logical Unit (SLU), has responsibility for controlling an LU-LU session and for such functions as error recovery.

propagation delay The time it takes for a signal to reach its destination. Assuming no induced delays (buffering at intermediate nodes, etc.), the propagation delay, in seconds, is approximately $3.3p \times 10^{-6}$, where p is the total path length, in kilometers. For a transatlantic data link (about 4000 km), the delay would be about 0.013 s for a cable link, or about 0.23 s for a satellite link (geostationary, orbiting at about 35,000 km).

protocol An agreement between parties on the format, meaning, and sequence of control messages to be exchanged between the parties. Protocols exist for device control, data link control, end-to-end data format control, and so on. The parties to a protocol may, for example, be corresponding SNA layers at either end of a data link.

protocol conversion Means by which a device or process can communicate with another device or process supporting a different protocol. Protocol conversion often involves built-in terminal emulation functions (e.g., when using an ASCII display terminal as an IBM 3270 terminal, via a protocol converter). IBM 3174 Establishment Controllers can perform 3270 protocol conversion for attached ASCII display terminals. See *IBM 3708, IBM 3710, IBM 5208,* and *IBM 5209* in *Sec. 3*.

protocol enveloping Process by which frames (message blocks) of one data link protocol carry, as information, the message blocks or other transmitted data (e.g., individual characters or groups of characters from start-stop terminals) of another data link protocol.

PSPDN Packet-switched public data network. See *packet switching*.

PTT Post, Telephone, Telegraph. Monopoly provider of postal, voice-grade telephone, and telegraph services. Many PTTs now include public data networks in their offerings. PTTs are common in Europe, but nonexistent in countries such as the United States and Canada. Many PTTs are being broken up, sometimes with the retention, for the resulting telecommunications company, of monopoly status (e.g., Deutsche Bundespost TELEKOM in Germany) and sometimes with a requirement to compete with alternative suppliers (e.g., British Telecom, Mercury, and the British cable TV suppliers).

PU See *Physical Unit*.

PUID Physical Unit Identifier. Low-order 20 bits (five hexadecimal digits) of an SNA station's ID. Transmitted in response to an XID command (see *SDLC frame*), along with the station's block number and Physical Unit type. IBM 3174 Establishment Controllers and 3274 Cluster Controllers are examples of SNA stations.

pulse-code modulation See *PCM* and *ADPCM*.

PU-PU session See *session*.

QAM Quadrature Amplitude Modulation. Modulation technique used by modems conforming to CCITT recommendation V.32, in which the signal is the sum of a sine wave and a cosine wave (90° apart), the amplitude of each of which is set to one of four discrete levels, depending on the value of the first or second pair of bits in a 4-bit group.

QLLC Qualified Logical Link Control. A facility, within ACF/NCP and SNA peripheral nodes, allowing SDLC data streams to be carried over X.25 packet networks.

RAS Reliability and serviceability. IBM marketing term, referring to their equipment maintenance goals and practices. Used in expressions such as "subsystem RAS tests."

requester See *ECF*.

response time The amount of elapsed time between the completion of a request (e.g., the depression of an AID key on an IBM 3270 terminal) and the receipt of the beginning of the reply.

reverse channel See *backward channel.*

reverse video Display attribute in which the foreground and background colors are exchanged (e.g., black character on green background, rather than green character on black background).

RFT See *DIA/DCA.*

RH Request/Response Header. In SNA, the RH is added to an RU (Request/Response Unit) by the Transmission Control layer. The BIU (Basic Information Unit) thus created is then passed to the Path Control layer. The RH is 3 bytes long. The high-order bit of the first byte indicates whether the RU is a request (0) or a response (1). See also *BIU, BTU, PIU, RU,* and *TH.*

RH	Request/Response Unit (RU)
Basic Information Unit (BIU)	

RJE (Remote Job Entry) See *JES.*

router Network node, which, alone or in tandem with other routers, ensures the delivery of a message from a station on one network to a station on another network. The sending station includes, in the message, the address of the receiving station and the address of the local router. The local router determines the best route to the receiving station, which may or may not involve intermediate routers. The sending and receiving station must use the same protocols at and above the Network layer (see *OSI Model*) but may use different data link protocols.

RU Request/Response Unit. In SNA, a general term used to describe units of information exchanged between Network-Addressable Units (NAUs). All such information is considered to be either a request or a response.

S3G See *3279 Model S3G* in *Sec. 4.*

SAA Systems Application Architecture. A set of IBM-defined standards designed to provide a consistent environment for programmers and users across a broad range of IBM equipment, including microcomputers, minicomputers, and mainframes. IBM has made a commitment to support SAA in both the SNA and OSI environments.

At the implementation level, SAA defines four major elements, namely Common Programming Interface (CPI), Common User Access (CUA), Common Applications, and Common Communications Support (CCS). The following diagram shows their relationship with one another and with other logical elements.

The Four Elements of SAA, as Depicted by IBM

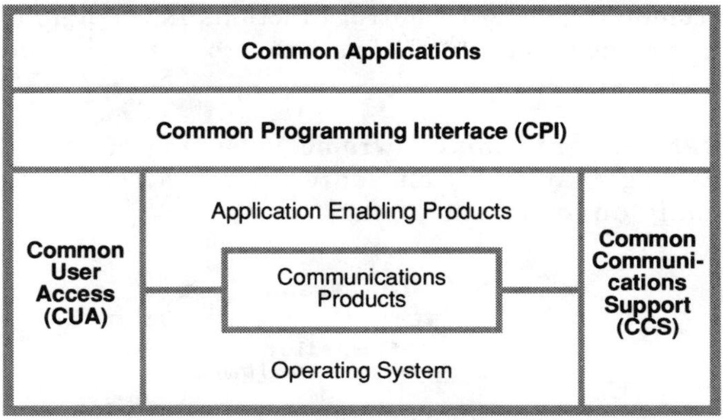

Common Applications

Common Programming Interface (CPI)

Common User Access (CUA)

Application Enabling Products

Communications Products

Operating System

Common Communi-cations Support (CCS)

SAA is one of three major approaches to the issue of providing a consistent environment. The others are common machine architecture (e.g., Digital Equipment Corporation's VAX series) and common operating system software (e.g., UNIX).

Sabre code Interchange code, based on BCD, defined for use in American Airlines' Sabre reservations system. In Sabre code, 16 of the codes used by BCD for punctuation and special characters are redefined as control codes.

scan code Code transmitted by the keyboard of a PC or a terminal each time a key is depressed or released. Each key's scan code is determined by its position rather than by the character engraved, embossed, or printed on it. Translation of keystrokes or keystroke combinations (e.g., Alt-A) into usable values (e.g., ASCII, IBM PC Extended Codes, 3270 buffer codes) is accomplished by software or microcode.

scrambler Within a DCE, a device or process which ensures, through "pseudo-randomizing," that an outgoing bit stream does not contain an excessive number of successive 1s or 0s. This protects against possible loss of synchronization. A descrambler performs the complementary function of restoring the bit stream to its original form. See also *NRZI*.

screen format See *IBM 3278* in *Sec. 3*.

SCS SNA Character String. See *SCS* in *Sec. 4*.

SDLC Synchronous data link control. IBM-defined link-control protocol which has the characteristics of being code-independent, being transparent to the bit pattern being handled, using a single format for the combination of data and control information, combining several control functions in a single transmission and being independent of path control. See also *synchronous* and *SDLC frame*.

SDLC frame A transmission frame consisting of beginning and ending flag characters, an address field, a control field, and optionally, an information field.

SDLC Frame

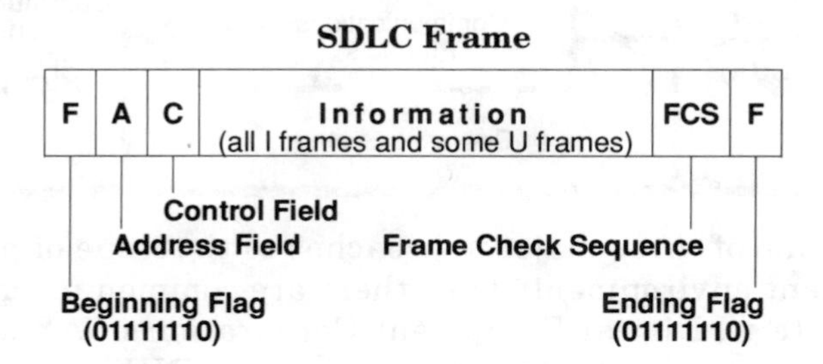

Each Flag is a unique sequence of 8 bits (01111110) which cannot occur anywhere else in the frame (see *zero-bit insertion*). Receiving stations use the Flags to recognize the beginning and end of a frame.

The Frame Check Sequence is used for error detection. It is generated by the transmitting station and regenerated for checking purposes by the receiving station. See *FCS* and *CRC*.

The Address Field is used for polling and addressing on multipoint data links. It is usually a single byte, but may be two or more bytes. It contains the address of a secondary link station sending or receiving the frame. For multi-byte Address Fields the low-order bit of all but the last byte is set to 1. The low-order bit of the last or only byte is set to 0. For point-to-point operation, the Address Field is set to 00000000.

There are three SDLC frame types, recognized by the low-order bit or low-order 2 bits of the first or only byte of the Control Field. Note that, in all SDLC frames, the low-order bit (bit 7) of each byte is transmitted first.

I Frame Information Frame. This frame is used for the transmission of a BTU (Basic Transmission Unit — see *Data Link Control,* under *SNA*). Its Control Field has the following format:

One-byte I Frame Control Field

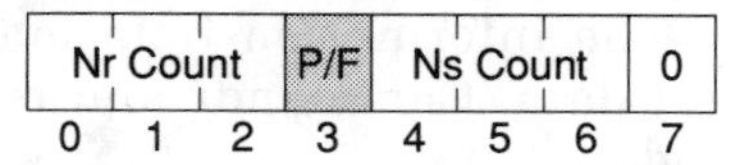

Two-byte I Frame Control Field

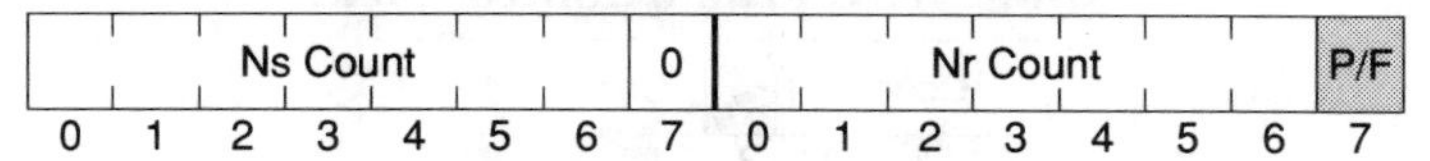

The Ns Count (sequence number) is incremented, modulo 8 (modulo 128 for 2-byte control fields), for each I Frame transmitted. The Nr count depends on the number of frames received since the link was initialized. It indicates the number of the next I Frame the station expects to receive.

S Frame Supervisory Frame. This frame is used for supervisory commands and responses, identified by a 2-bit code. It does not contain an information field.

One-byte S Frame Control Field

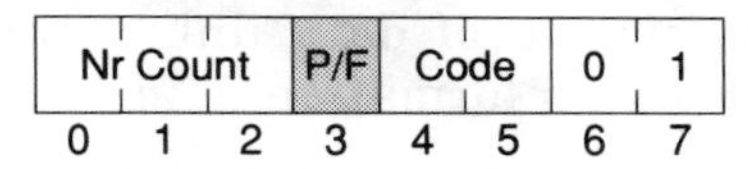

Two-byte S Frame Control Field

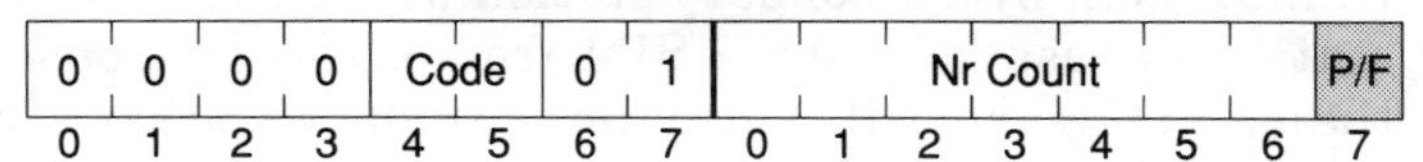

There are three possible codes.

00—RR (Receiver Ready) is used by a primary station for polling and by a secondary station to indicate that it is ready to receive more I Frames. The Nr count confirms the last received frame sequence number.

01—RNR (Receiver Not Ready) is used by a primary or secondary station to indicate that it is not ready to receive more I Frames, because of lack of buffer space.

10—REJ (Reject) is used by a receiving station to indicate a frame sequence error to the sending station. The Nr count indicates the frame number at which retransmission is to begin and implicitly confirms satisfactory receipt of frame Nr–1.

U Frame Unnumbered Frame. This frame is used for data link management, which includes activation, initialization, and response mode control of secondary stations. It also includes the reporting of procedural errors that are not recoverable by retransmission. The information field may be used for data link management data. Commands and responses are identified by a 5-bit code.

One-byte U Frame Control Field

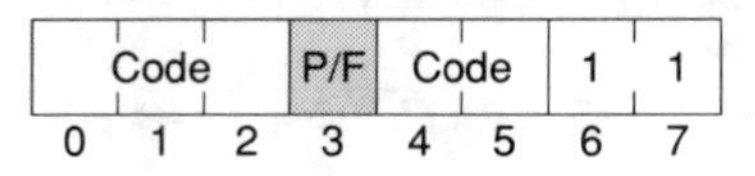

Two-byte U Frame Control Field

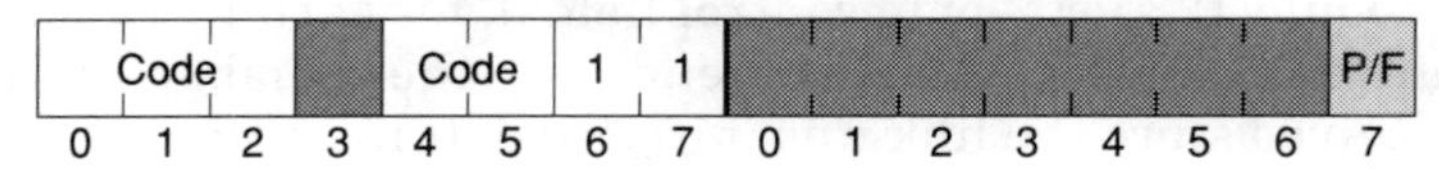

The 5-bit code field may be used for commands (from a primary station) or responses (from secondary stations). Several command/response descriptions follow.

000 00—UI Unnumbered Information (command or response). A UI frame is used as a command or a response for the transmission of data link management data. The frame includes an information field.

000 01—RIM Request Initialization Mode (response). A RIM frame is sent by a secondary station. It requires the primary station to respond with a SIM frame. Any response other than a SIM frame will cause the secondary station to resend the RIM frame. The F bit is set to 1.

000 01—SIM Set Initialization Mode (command). This command is sent by a primary station to begin initialization procedures. Both stations will set their Ns and Nr counts for the link to 0. The expected response from the secondary station is a UA frame. The P bit is set to 1.

000 11—DM Disconnect Mode (response). This response is sent by a secondary station that is in normal disconnected mode (NDM), if it receives a command from the primary station other than SIM, SNRM, or TEST.

001 00—UP Unnumbered Poll (command). Optional-response poll sent by the primary station on a multipoint or loop link. The Address Field indicates the station being polled. A P bit of 0 only invites a response, but a P bit of 1 requires a response. An affirmative response consists of one or more I Frames.

010 00—DISC Disconnect (command). This command is sent

by the primary station to place the secondary station off line. The secondary station will remain disconnected until it receives an SNRM or SIM command. The P bit is set to 1.

011 00—UA Unnumbered Acknowledgment (response). Used as an affirmative response to SIM, DISC, or SNRM. The F bit is set to 1.

100 00—SNRM Set Normal Response Mode (command). Used by the primary station to place the secondary station in a subordinate mode, which will not allow it to make unsolicited transmissions. The expected response is a UA frame. Both stations will set their Ns and Nr counts for the link to 0. The P bit is set to 1.

100 01—FRMR Frame Reject (response). Sent by a secondary station in normal response mode (NRM) when it receives a nonvalid command. A command may be nonvalid because it is not implemented at the secondary station, because the information field is too long or not allowed, or because the command's Nr field is incongruous with respect to the last Ns field sent by the secondary station. The F bit is set to 1.

101 11—XID Exchange Station Identification (command or response). Sent by a primary station, typically (but not necessarily) on a switched (dial-up) link, to solicit the secondary station's identification. Used as a response, by the secondary station, to send block number, physical unit type, and identification information (see *PUID*), which usually requires a 4-byte information field. The P/F bit is set to 1 in both cases.

110 00—CFGR Configure (command or response). Used to perform loop configuration functions. Subcommands, in the information field, are used to control loop operation.

111 00—TEST Test (command or response). Used by the primary station to request a test response from a secondary station, regardless of the secondary station's mode. In its response, the secondary station will include whatever information field was sent by the primary station (unless it has insufficient buffer capacity). The P/F bit is set to 1 in both cases.

Poll/Final (P/F) bit This bit is used for send/receive control. It has a value of 1 for polling by a primary station and a secondary station's response to polling. For I frames, it is set to 1 to indicate the last of a sequence of frames. Otherwise it is set to 0. When used in a frame sent by a primary station, it is called the P bit. When used in a frame sent by a secondary station, it is called the F bit.

secondary channel See *backward channel*.

Secondary Logical Unit (SLU) Typically, an LU Type 1, 2, 3, 4, or 7 terminal. A Secondary Logical Unit usually has less power than (i.e., is subordinate to) a Primary Logical Unit, and it does not normally bear responsibility for control of a session or for error recovery.

Server See *ECF*.

Service Point See *NetView*.

session A temporary logical connection between two NAUs in an SNA network. In a hierarchical SNA network, there are five possible NAU-NAU session types. A sixth session type exists in APPN (Advanced Peer-to-Peer Networking).

Type	Description
CP-CP (APPN only)	Control-Point-to-Control-Point session, in APPN, always established in parallel pairs—one send and one receive. Used in the exchange of network information among adjacent network nodes.
LU-LU	Session in which communication takes place between two Logical Units or, for example, between an LU Type 2 and a host application. Except for sessions involving LU Type 6.0, 6.1, or 6.2, LU-LU sessions are always established via the SSCP.
PU-PU	Session used to allow one Physical Unit (e.g., subarea node) to inform another of a physical event, such as a path interruption.
SSCP-LU	Connection between a Logical Unit [e.g., LU Type 2 (keyboard/display)] and the System Services Control Point, prior to the establishment (binding) or following the termination (unbinding) of an LU-LU session. Initiated by an ACTLU (Activate Logical Unit) request. Terminated by a DACTLU (Deactivate Logical Unit) request.
SSCP-PU	Connection between a Physical Unit [e.g., PU Type 2 (cluster or establishment controller)] and the System Services Control Point, prior to the establishment or following the termination of an SSCP-LU session. Initiated by an ACTPU (Activate Physical Unit) request. Terminated by a DACTPU (Deactivate Physical Unit) request.
SSCP-SSCP	Used for cross-domain network services. For example, where a secondary logical unit in one domain needs to establish a session with a primary logical unit in another domain, an SSCP-SSCP session would be used to exchange path and address information between the two domains. SSCP-SSCP sessions are established at network initialization time and remain established as long as cross-domain LU-LU sessions are permitted.

See also *logical unit, NAU, physical unit,* and *SSCP*.

Session layer See *OSI*.

shift in Control code used, in data communications, to indicate a return to the default meaning for subsequent data codes.

shift out Control code used, in data communications, to indicate an alternative meaning for subsequent data codes.

signal constellation See *signal space diagram.*

signal ground Connection providing a common return for all other circuits (except frame ground) in an unbalanced DTE-to-DCE interface. See also *CCITT Recommendation V.24* in *Sec. 2.*

signal space diagram A diagram depicting bit or bit-group values in positions which reflect the relative phase angles and amplitudes of their modulated form. The following, very simple diagram depicts the relative phase angles of the four possible dibit (bit-pair) values, as modulated according to *CCITT Recommendation V.22*, which specifies differential phase-shift keying. Because no amplitude modulation is involved, all four points are equidistant from the center.

CCITT V.22 Modem Signal Space Diagram

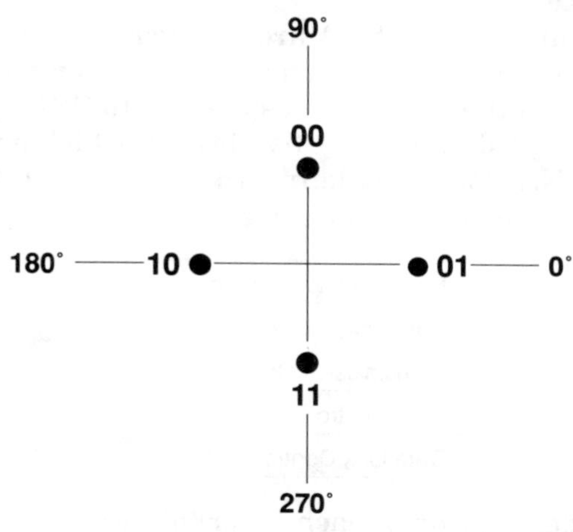

single-session Of a terminal or terminal-emulation device, having access to only one host session at a time.

single-station Of a terminal emulation product, having a host connection other than through a gateway.

Six-bit Transcode A rarely used code associated with the BSC protocol, intended for the transparent transmission of data.

Each byte has 6 significant low-order bits and 2 high-order bits, both of which are set to 1. As BSC control codes have a value of 0 in the high-order bit, data bytes can never be mistaken for control codes. The CRC (cyclic redundancy check) field of a Six-bit Transcode frame is 12 bits long.

SMTP See *TCP/IP*.

SNA Systems Network Architecture. IBM network architecture, defined in terms of its functions, formats, and protocols. In *Communications Architecture for Distributed Systems,* R. J. Cypser states that SNA can be understood in terms of six concepts, specifically:

- Network-addressable units (NAUs)
- Sessions
- Functional layers (see below)
- Function subsets within each layer
- Peer protocols
- Control domains (see *SSCP*)

The functional layers of SNA are somewhat similar to those of the OSI Model, with some variations in layer boundaries and nomenclature in the layers corresponding to OSI layers 3, 4, and 5. Note that SNA defines only five layers, which are not usually numbered. SNA does not define the application and physical control layers. The SNA layers are:

Function Management
Data Flow Control
Transmission Control
Path Control
Data Link Control

A summary description of each layer follows.

Function Management Interprets and/or modifies data entering or leaving each End User. Formats, edits, or maps data for various presentation mechanisms and coordinates resource allocation. Provides for application services. For outgoing Request/Response units (RUs), adds control information and passes them to the Data Flow Control layer. Deals with incoming RUs according to the control information they contain.

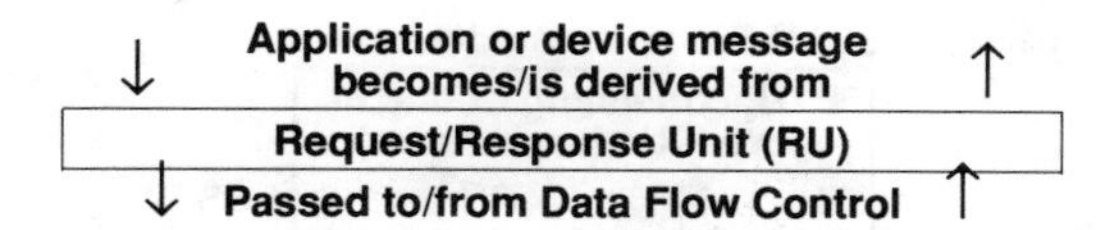

Data Flow Control Regulates the flow of data between the end points of a session and ensures flow integrity. (Does not control rate of flow.) May create its own RUs for control purposes and may, in turn, respond to incoming control RUs. Passes RUs from the Function Management layer directly to the Transmission Control layer (and vice versa).

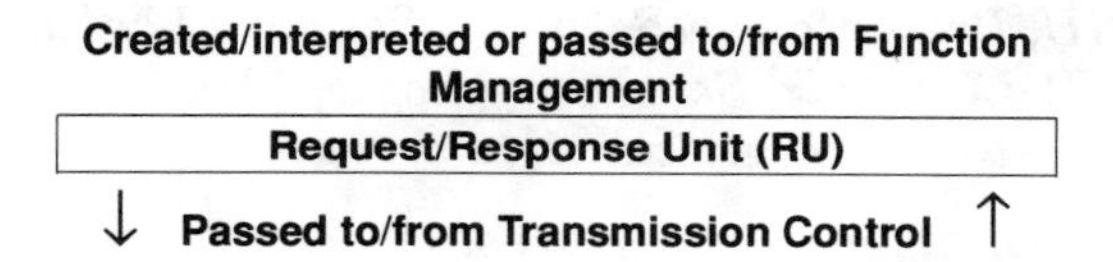

Transmission Control Coordinates each session's transmissions, including sequence numbering and rate control. Operates at each end of a session. Adds a Request/Response Header (RH) to RUs passed from the Data Flow Control layer, creating Basic Information Units (BIUs), which it passes to the Path Control layer. Removes the RH from incoming BIUs.

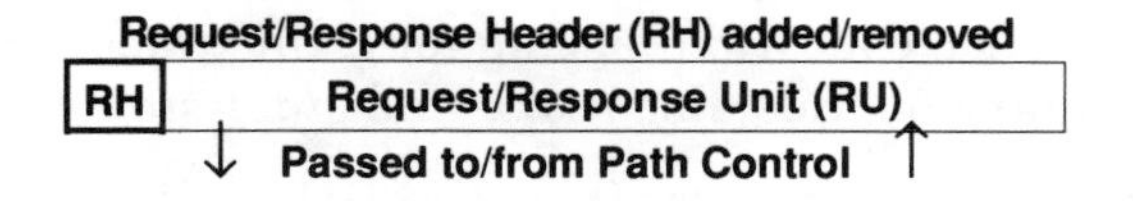

Path Control Routes all messages to links and destinations. Adds a Transmission Header (TH) to BIUs passed from the Transmission Control layer, creating Path Information Units (PIUs) and Basic Transmission Units (BTUs, containing one or more PIUs), which it passes to the Data Link Control layer. Removes the TH from incoming PIUs.

Transmission Header (TH) added/removed

TH	RH	Request/Response Unit (RU)

↓ Passed to/from Data Link Control ↑

Data Link Control Controls the flow of data on the link. For SDLC, adds beginning and ending flag characters, address field, and control field to BTUs passed from the Path Control Layer, creating SDLC frames. For incoming SDLC frames, removes the flags and the address and control fields.

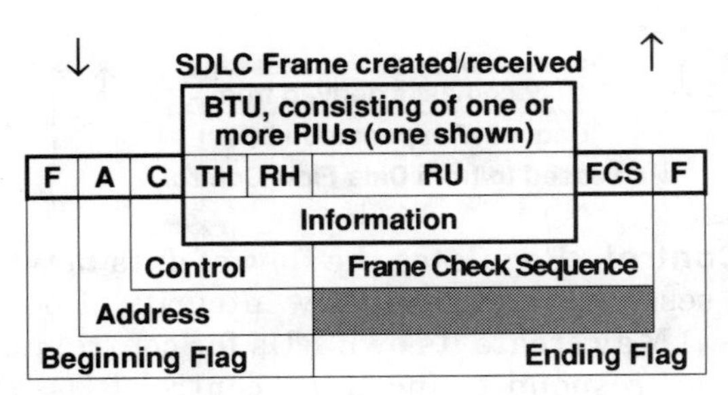

For further SNA definitions, see *address, BIND, binding, BIU, bracket, BTU, chain, control field, explicit route, flag field, frame, half session, information field, Logical Unit, NAU, Physical Unit, PIU, Presentation Services, Primary Logical Unit, RH, RU, SDLC, SDLC frame, Secondary Logical Unit, session, SSCP, TH, UNBIND, unbinding,* and *virtual route.*

SNA/DS SNA Distribution Services. A set of SNA transaction programs that makes use of LU Type 6.2 and APPC services to provide a store-and-forward message delivery service. Also called *SNADS.*

SOH Start Of Header. BSC control character used to signify that header information follows. Header information may relate to message routing or to device-dependent functions.

space For an interchange circuit, a condition representing a value of 0. A space condition lasting for *n* bit times represents *n* successive binary 0s. See also *mark* and *start-stop.*

SPF Structured Programming Facility. Predecessor of *ISPF/PDF.*

split-screen 1. Display terminal function allowing the concurrent display of two (usually) or more discontiguous portions of a common presentation space. (Not to be confused with dual- or multi-session displays in which part or all of each of several presentation spaces is displayed.)

2. Application function providing a display terminal operator with two (usually) or more partial-screen displays, each corresponding to a different application task. IBM's ISPF is one of the better-known applications offering a split-screen function.

(Many noncommunicating applications, such as PC-based word processing programs, also support split-screen operation.)

SRPI Server/Requester Program Interface. Application programming interface defined by IBM for use with ECF in conjunction with LU Type 2. See also *ECF* and *Logical Unit.*

SSCP System Services Control Point. Host-based NAU (Network Addressable Unit) through which all LU-LU sessions (other than those between LU Type 6.0s, 6.1s, or 6.2s) are established. The portion of the network controlled by an SSCP is called a domain. See also *session*.

SSCP-LU session See *session*.

SSCP-PU session See *session*.

SSCP-SSCP session See *session*.

start-stop A data transmission mode in which synchronization is established for each character. Each character is preceded by a "start" bit and followed by at least one "stop" bit. The absence of transmitted data is indicated by a continuous "mark" condition (continuous 1s). A start bit, which is a "space" condition (a zero) for the duration of 1 bit time (e.g., 1/1200 s at 1200 bit/s) indicates that a string of bits representing a transmitted character follows (from 5 to 9 bits, depending on the code and equipment being used). A resumed mark condition for at least 1 bit time after the final character bit (up to at least 2 bit times for some older equipment) is referred to as a stop bit or bits. Start-stop is often referred to as asynchronous transmission.

station A physical input or output point in a communication system, including intermediate terminals (e.g., cluster controller) and end terminals (e.g., IBM 3278).

station address See *address*.

status line Nondata line on a display terminal, used to display error messages, communication status, keyboard status, etc.

store and forward Descriptive of data communications applications (e.g., electronic mail) in which messages are stored at an intermediate point until the receiver is ready to retrieve them, or until a communication link to the receiver is available. See also *message switching*.

STX Start Of Text. BSC control character used to signify that text characters follow.

subarea A group of addressable elements in an SNA network having the same subarea ID.

subarea ID A subfield of an SNA network address. See also *address*.

subarea node A node that is able to communicate directly with peripheral nodes and with other subarea nodes. See also *physical unit* (*PU Type 4* and *PU Type 5*).

SWIFT Society for Worldwide Interbank Financial Telecommunications. International organization of banks, under whose ægis a network is operated for the transfer of funds and the exchange of financial information.

switched line See *dial-up line*.

switched virtual circuit See *virtual circuit*.

SYN Synchronous idle. Eight-bit character, with the value 00110010 (EBCDIC) or 00010110 (ASCII), used for data synchronization in Binary Synchronous Communication (BSC). See also *synchronous*.

sync Abbreviation of synchronous.

synchronous 1. A data transmission mode in which synchronization is established for an entire block (frame) of data.

In Binary Synchronous Communication (BSC), a special control character (SYN) is used to allow the receiving device to establish character synchronization (i.e., identify character boundaries within the string of bits). Pairs of SYN characters, which are inserted by the transmitting device and removed by the receiving device, may appear at the beginning of a block or within a long block. In transparent mode, SYN characters (and other control characters) not intended for receipt as data are preceded by a Data Link Escape character (DLE).

In Synchronous Data Link Control (SDLC) and High-Level Data Link Control (HDLC), a special 8-bit sequence (01111110), called a flag field, is transmitted at the start and end of every block of data. The beginning flag field ensures bit synchronization and establishes byte synchronization for the address field and control field.

Within transmitted data, SDLC/HDLC devices insert a 0 bit after every occurrence of five successive 1 bits. On receipt, a 0 bit, which occurs within a message block and following five successive 1 bits, is deleted. This ensures the occurrence of a 0 bit at least every six bit times, which, in conjunction with NRZI encoding, guarantees the maintenance of bit synchronization. Also, data bytes with a value of 01111110 are not confused with flag fields, which are inserted into and removed from the data stream at the DTE-to-DCE interface.

2. The operational mode of a class of modems and similar devices.

See *isochronous*.

T carrier A digital transmission facility, using time-division multiplexing and operating at one of several defined frequencies in the range 1.544 to 274.176 Mbit/s.

T1 1.544-Mbit/s facility, used to transmit eight thousand 193-bit frames per second (8000 x 193 = 1,544,000). Each 193-bit frame is made up of twenty-four 8-bit groups (octets) and a framing bit. Each octet is derived, by time-division multiplexing, from one of twenty-four 64-kbit/s channels, each of which may be carrying data or voice traffic. For voice traffic, the octets are derived by sampling the analogue (voice) signal 8000 times per second for any of 256 amplitude levels [a technique referred to as pulse code modulation (PCM)].

In Europe, a similar facility, operating at 2.048 Mbit/s and loosely referred to as a CEPT carrier (see *CEPT*), is used to transmit information from thirty 64-kbit/s data or voice channels, plus two channels containing signalling and framing information.

T1C 3.152-Mbit/s facility, multiplexing forty-eight 64-kbit/s channels (2xT1).

T2 6.312-Mbit/s facility, multiplexing ninety-six 64-kbit/s channels (4xT1 or 2xT1C).

T3 44.736-Mbit/s facility, multiplexing six hundred and seventy-two 64-kbit/s channels (7xT2).

T4 274.176-Mbit/s facility, multiplexing four thousand and thirty-two 64-kbit/s channels (6xT3).

TCAM See *ACF/TCAM*.

TCP/IP Transmission Control Protocol/Internet Protocol. Protocol introduced, in the early 1970s, by the U.S. Department of Defense (DOD) for the interconnection of DOD networks. It has become a de facto standard for many LAN-to-host and LAN-to-LAN connections. TCP is used to establish and maintain sessions between users and supports an application program interface (API). IP provides for network interconnection and uses global source and destination addressing schemes. In addition to TCP and IP, the TCP/IP protocol suite includes the following protocols.

FTP File Transfer Protocol. Used for the transfer of data files consisting of one or more segments.

ICMP Internet Control Message Protocol. A part of IP, used for the forwarding, to a control point, of status messages generated by internetwork gateways.

SMTP Simple Mail Transfer Protocol. Used for electronic mail transfer between TCP/IP users.

Telnet Application-level protocol that makes a remote terminal emulator appear, to a host system, to be a local terminal.

UDP User Datagram Protocol. Used for the transmission of messages without end-to-end error checking or session management.

See also *Internet*.

TDM systems Time-division multiplex systems. See *T carrier*.

Telnet See *TCP/IP*.

TH Transmission Header. In SNA, the component of a BTU (Basic Transmission Unit) that is used to guide the BTU, over various communication links, to its destination. It can be 2, 6, 8, or 24 bytes long and is added to the BIU (Basic Information Unit) by the Path Control layer.

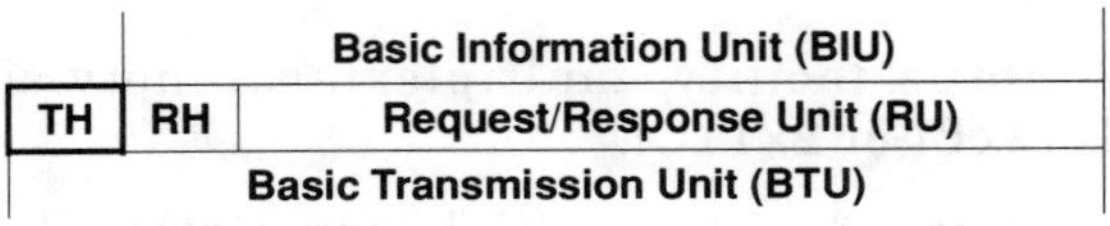

	Basic Information Unit (BIU)	
TH	RH	Request/Response Unit (RU)
Basic Transmission Unit (BTU)		

The first 4 bits of a Transmission Header contain a Format Identifier (FID). The following FID types are defined:

TH Types, by Format Identifier (FID)

FID	Length (bytes)	Used for ...
FID0	8	Messages passed between an SNA node and a non-SNA device
FID1	8	Messages passed between subarea nodes where one (or both) of the nodes does not support explicit route and virtual route processing
FID2	6	Messages passed between a subarea node and a PU Type 2 peripheral node
FID3	2	Messages passed between a subarea node and a PU Type 1 peripheral node
FID4	24	Messages passed between subarea nodes where both nodes support explicit route and virtual route processing
FIDF	24	Special messages related to message sequencing

See also *BIU, BTU, PIU, RH,* and *RU*.

throughput Net data transfer rate between an information source and an information sink (destination). The American

National Standards Institute (ANSI) defines net throughput TRIB (Transfer Rate of Information Bits) as being the number of information bits accepted by the sink divided by the total time required to get those bits accepted.

TRIB can be calculated with the following formula:

$$\text{TRIB} = \frac{B(M-C)(1-p)}{t + \dfrac{BM}{R}} \text{ bit/s}$$

where B is the number of bits per character, M is the block length, in characters, C is the number of noninformation characters, p is the probability of an error occuring in a block, t is the time between blocks (in seconds), and R is the nominal data rate, in bits per second.

Even for advanced link protocols like SDLC, throughput rarely exceeds 80 percent of the nominal link data rate (e.g., effective rate of about 3750 bit/s over a 4800 bit/s full-duplex link). However, data compression techniques can produce an effective data rate exceeding the nominal rate (see *MNP* and, in *Sec. 2, CCITT Recommendation V.42 bis*). See also *CCITT Recommendation I.122* in *Sec. 2.*

training Process through which isochronous ("synchronous") modems achieve synchronization with one another.

transaction Sequence of events involving the entry and acceptance of an arbitrary set of data.

transaction-intensive Descriptive of applications in which the rate of completion of transactions is very high. Terminal-based applications that are transaction-intensive tend to require very fast response times. See *data entry*.

Transmission Control See *SNA*.

transparent mode Mode of transmission in which all information in a message is to be treated as data, except where a special "escape" character is used. See also *DLE* and *synchronous*.

Transport layer See *OSI Model*.

transport mechanism Generic term for Session-layer software components. See *OSI Model* for Session layer description.

trellis coding Advanced modulation technique, characterized by low noise sensitivity and forward error-correction (FEC) capability. *CCITT recommendation V.32* specifies trellis coding as an option.

TRIB Transfer Rate of Information Bits. See *throughput*.

TSO Time Sharing Option. User terminal support under any of the variants of MVS or VM, providing a command language, access to individual and shared data sets (files), a structured editing and data set management facility (ISPF), utilities, compilers, etc. Access to host resources is allocated on a time-sliced basis.

TTP Telephone Twisted Pair. See *twisted-pair cable*.

turnaround time See *line turnaround*.

twisted-pair cable Cable consisting of one or more parallel sets of two insulated wires twisted together. The two most common types are telephone twisted pair (TTP) and the IBM Cabling System.

UDP See *TCP/IP*.

UDLC Universal Data Link Control. Sperry-Univac (now Unisys) data link protocol, based on but not compatible with HDLC.

unbalanced circuit Circuit (in a DTE-to-DCE interface, for example) which makes use of a common return circuit (signal ground). Groups of unbalanced circuits require fewer conductors than their balanced counterparts ($n + 1$, rather than $2n$, where n is the number of circuits), but are sensitive to differences in ground potential. See also *CCITT Recommendations V.10* and *V.28* in *Sec. 2*.

UNBIND Request sent from one NAU to another (e.g., a primary LU to a secondary LU) to deactivate the session between them. See *BIND*.

unbinding The process, in SNA, by which a session between NAUs is terminated. See *binding*.

Unicode Universal interchange code. Sixteen-bit interchange code being developed by a consortium of computer manufacturers and intended to eventually replace codes such as ASCII and EBCDIC. With its 65,536 possible values, Unicode is expected by its proponents to cover, without ambiguity, every useful letter (Roman, Greek, Arabic, Hebrew, Cyrillic, etc., with all valid letter and diacritical mark combinations), symbol, punctuation character, and ideographic character (Chinese, Japanese, Korean, etc.).

uploading Workstation-to-host file transfer.

vector graphics See *vector graphics* in *Sec. 4*.

virtual circuit Circuit used in packet networks and defined solely by the addresses of the end points rather than by the physical path used to communicate between them. The physical path, segments of which or all of which can be shared with other virtual circuits, varies from call to call, depending on existing traffic on the possible paths. The path used by a virtual circuit can vary during a call. Each data packet transmitted over a virtual circuit contains the virtual circuit number (origin and destination address pair).

Virtual circuits may be switched or permanent. Switched virtual circuits are established by the exchange between DTEs and the network of call-setup packets and removed by the exchange of call-clearing packets. Permanent virtual circuits do not require setup or clearing. See also *packet switching*.

virtual route SNA term defining the logical connection that appears to exist between subarea nodes. Virtual routes are assigned to explicit routes, each of which may be used by one or more virtual routes.

VM/370 Virtual Machine/370. IBM operating system which controls the concurrent operation of multiple virtual machines on an IBM System/370 or System/390. Each virtual machine may be controlled by its own operating system (e.g., MVS, VSE). Implementations of VM/370 include VM/XA (Extended Architecture) and VM/ESA (Enterprise Systems Architecture).

VM/CMS Virtual Machine/Conversational Monitor System. IBM host software characterized by the fact that each user has his or her own "virtual" host machine, along with a command language and a number of associated applications and utilities (e.g., text editors). The Virtual Machine Facility may also be used to run one or more copies of other operating systems, such as MVS and VSE.

VMS Operating system for Digital Equipment Corporation's VAX series of computers.

voice-grade circuit Switched (dial-up) or leased (dedicated) telephone circuit, capable of carrying analogue signals with frequencies between 300 and 3300 Hz.

VRC Vertical redundancy check. Error checking technique which includes an extra (redundant) bit in each character being transmitted by a DTE. The extra bit is set to 0 or 1 in order to make the total number of 1 bits either even (for even parity) or odd (for odd parity).

VSE Virtual Storage Extended. IBM operating system commonly used on smaller IBM System/370 and System/390 host systems. Formerly called DOS/VSE (Disk Operating System/Virtual Storage Extended). The latest implementation of VSE is VSE/ESA (Enterprise Systems Architecture).

VT52 An ASCII display terminal formerly manufactured by DEC (Digital Equipment Corporation). Along with the VT100, one of the terminals most commonly emulated on personal computers.

VT100 An ASCII display terminal formerly manufactured by DEC, which was the best-known terminal to support control sequences (escape sequences) defined by ANSI. In terminal emulation, VT100 is used as a generic term to denote full ANSI X3.64 conformance.

VT200 A series of ASCII display terminals formerly manufactured by DEC. Includes the VT220 and the VT241 color display.

VT300 A series of high-resolution ASCII terminals manufactured by DEC. Includes the VT320 text terminal, the VT330 monochrome text and graphics terminal, and the VT340 color text and graphics terminal. The VT330 and VT340 can handle dual sessions. All three terminals operate at up to 19,200 bit/s.

VT420 High-resolution, dual-session ASCII text terminal manufactured by DEC. Operates at up to 38,400 bit/s, allows viewing of one or both sessions and provides a display area with 24, 25, 36, or 48 lines of text, with 80 or 132 characters per line.

VTAM See *ACF / VTAM*.

VTAME See *ACF / VTAME*.

windowing Display technique in which multiple sessions or multiple application displays can be viewed concurrently within rectangular, bordered areas on the screen.

workstation Contemporary term for a personal computer or terminal device, usually but not necessarily operating within a local area network, which is used by someone to perform the greater part of his or her everyday work. At one time expanded to "integrated workstation," because of the integration of a number of functions (e.g., spreadsheet accounting, database access, host access, word processing) within or at the same device. Alternatively written as two words — *work station*.

XID Exchange station identification. SNA command or link response used to verify the identity of the nodes being connected on a switched network (e.g., dial-up line). (May also be used for nonswitched lines.) See also *PUID.*

zero-bit deletion Process, in HDLC and SDLC operation, whereby a 0 bit is removed from a received data stream every time it occurs immediately after five successive 1 bits. This is the complementary function to zero-bit insertion.

zero-bit insertion Process, in HDLC and SDLC operation, whereby a 0 bit is inserted into a data stream, as it is transmitted, every time five successive 1 bits are encountered. This ensures that flag fields are not treated as data and that, within the data stream, a zero bit occurs at least once every 6 bits. See also *NRZI* and *synchronous.*

zero-complemented differential encoding See *NRZI.*

Section

2

Recommendations and Standards

Introduction

It would be impossible in a book (or even 50 books) of this size to provide a full explanation of every recommendation or standard applicable to data communications. However, given the importance of standards, it would be negligent to omit them altogether.

This section does the following:

- Identifies all ANSI, CCITT, EIA, IEEE, and ISO standards and Recommendations relating to data communications, including the full wording of the title of each one. (Note that CCITT always capitalizes the word Recommendations.)
- Provides an abstract of some of the CCITT Recommendations and of a number of standards.
- Provides a summary of some of the better-established Recommendations and standards, including the ISO interface connector standards.
- Enhances the usefulness of this book as a companion volume to textbooks on various data communications topics, especially those that deal with OSI, ISDN, and other emerging standards.
- Directs the reader to a number of sources of information on both national and international standards.

This section does not include:

- National standards for countries other than the United States.
- Regional standards, such as those defined by ECMA (European Computer Manufacturers' Association).

It does, however, provide the names, addresses, and telephone, fax and/or telex numbers of all national organizations that are members or corresponding members of ISO.

AMERICAN NATIONAL STANDARDS INSTITUTE (ANSI) STANDARDS FOR DATA COMMUNICATIONS

The American National Standards Institute does not develop standards. Rather, its role is that of a national coordinator of voluntary standards activities and as an approval organization and clearing house for consensus standards.

ANSI is privately funded. It is a member organization within ISO (International Organization for Standardization) and serves as the U.S. source for information on ISO standards.

ANSI uses its own numbering scheme where a sponsor does not supply a designation or if the standard was developed by an accredited standards committee. All ANSI numbers include the year of the standard and, where applicable, a revision year in parentheses.

Where an ANSI standard corresponds to a CCITT Recommendation, the CCITT number is given in brackets.

Data Communications and Related Standards Within the T1 Series (Telecommunications)

ANSI T1.101-1987 Telecommunications—Synchronous Interface Standards for Digital Networks.

ANSI T1.102-1987 Telecommunications—Digital Hierarchy—Electrical interfaces.

ANSI T1.103-1987 Telecommunications—Digital Hierarchy—Synchronous DS3 Format Specifications.

ANSI T1.103a-1990 Telecommunications—Digital Hierarchy—Synchronous DS3 Format Specifications.

ANSI T1.105-1988 Telecommunications—Digital Hierarchy—Optical Interface Rates and Format Specifications.

ANSI T1.106-1988 Telecommunications—Digital Hierarchy—Optical Interface Specifications (Single Mode).

ANSI T1.107-1988 Telecommunications—Digital Hierarchy—Format Specifications.

ANSI T1.107a-1990 Telecommunications—Digital Hierarchy—Supplement to Format Specifications (DS3 Format Applications).

ANSI T1.110-1987 [CCITT Q.700] Telecommunications—Signalling System Number 7 (SS7)—General Information.

ANSI T1.111-1987 [CCITT Q.701] Telecommunications—Signalling System Number 7—Functional Description of the Signalling System Message Transfer Part.

ANSI T1.112-1988 [CCITT Q.711-Q.716] Telecommunications—Signalling System Number 7—Signalling Connection.

ANSI T1.113-1988 [CCITT Q.761-Q.766] Telecommunications—Signalling System Number 7 — Integrated Services Digital Network (ISDN) User Part.

ANSI T1.114-1988 [CCITT Q.771-Q.775] Telecommunications—Signalling System Number 7—Transaction Capability Application Part (TCAP).

ANSI T1.115-1989 Telecommunications—Monitoring and Measurements for Signalling System Number 7 Networks.

ANSI T1.116-1989 Telecommunications—Signalling System Number 7—Operations, Maintenance, and Administration Part (OMAP).

ANSI T1.219-1991 Telecommunications—Integrated Services Digital Network (ISDN) Management—Overview and Principles.

ANSI T1.308-1990 Telecommunications—Integrated Services Digital Network (ISDN) Primary Rate—Customer Installation Metallic Interfaces Layer 1 Specification.

ANSI T1.601-1988 Telecommunications—Integrated Services Digital Network (ISDN)—Basic Access Interface for Use on Metallic Loops for Application on the Network Side of the NT (Layer 1 Specification).

ANSI T1.602-1989 Telecommunications—Integrated Services Digital Network (ISDN)—Data-Link Layer Signalling Specification for Application at the User-Network Interface.

ANSI T1.603-1990 Minimal set of Bearer Services for the ISDN Primary Rate Interface (PRI).

ANSI T1.604-1990 Minimal set of Bearer Services for the ISDN Basic Rate Interface (BRI).

ANSI T1.605-1989 Telecommunications—Integrated Services Digital Network (ISDN)—Basic Access Interface for S and T Reference Points (Layer 1 Specification).

ANSI T1.606-1990 Integrated Services Digital Network (ISDN)—Architectural Framework and Service Description for Frame-Relaying Bearer Service.

ANSI T1.607-1990 Integrated Services Digital Network (ISDN)—Layer 3 Signalling Specification for Circuit-Switched Bearer Service for Digital Subscriber Signalling System Number 1.

ANSI T1.609-1990 Interworking between the ISDN User-Network Interwork Interface Protocol and the Signalling System Number 7 ISDN User Part.

ANSI T1.610-1990 Telecommunications—Digital Subscriber Signalling System Number 1 (DSS1)—Generic Procedures for the Control of ISDN Supplementary Services.

ANSI T1.611-1991 Telecommunications—Signalling System Number 7 (SS7)—Supplementary Services for non-ISDN subscribers.

ANSI T1.613-1991 Telecommunications—Integrated Services Digital Network (ISDN)—Call Waiting Supplementary Service.

Data Communications Standards Within the X3 Series (Computers and Data Communications)

ANSI X3.1-1987 Information Systems—Data Transmission—Synchronous Signalling Rates. This standard is a minor variant of a combination of CCITT Recommendations V.6 and V.7.

ANSI X3.4-1986 Coded Character Set—7-bit American National Standard Code for Information Interchange (ASCII). This is the U.S. variant of CCITT International Alphabet Number 5 (CCITT Recommendation T.50, ISO 646:1983). See *Sec. 5, Interchange Codes*.

ANSI X3.6-1965(R1983) Perforated Tape Code for Information Interchange.

ANSI X3.15-1976(R1983) Bit sequencing of the American National Standard Code for Information Interchange in Serial-by-Bit Data Transmission.

ANSI X3.16-1976(R1983) Character Structure and Character Parity Sense for Serial-by-Bit Data Communication in the American National Standard Code for Information Interchange.

ANSI X3.25-1976(R1983) Character Structure and Character Parity Sense for Parallel-by-Bit Data Communication in the American National Standard Code for Information Interchange.

ANSI X3.28-1976(R1986) Procedures for the Use of the Communications Control Characters of American National Standard Code for Information Interchange in Specified Data Communication Links.

ANSI X3.32-1990 Graphic Representation of the Control Characters of ASCII.

ANSI X3.41-1990 Code Extension Techniques for Use with the 7-bit Coded Character Set of ASCII.

ANSI X3.44-1990 Determination of the Performance of Data Communications Systems.

ANSI X3.57-1977(R1986) Structure for Formatting Message Headings for Information Interchange Using the American National Standard Code for Information Interchange for Data Communication System Control.

ANSI X3.64-1979(R1990) Additional Controls for Use with the American National Standard Code for Information Interchange.

ANSI X3.66-1979(R1990) Advanced Data Communication Control Procedure (ADCCP).

ANSI X3.79-1981 Determination of Performance of Data Communications Systems that Use Bit-Oriented Communications Control Procedures.

ANSI X3.92-1981(R1987) Data Encryption Algorithm.

ANSI X3.100-1989 Information systems—Interface Between Data Terminal Equipment (DTE) and Data Circuit-terminating Equipment (DCE) for Operation with Packet-Switched Data Communications Networks (PSDNs), or Between two DTEs, by Dedicated Circuit.

ANSI X3.102-1983(R1990) Data Communication User-Oriented Performance Parameters.

ANSI X3.105-1983(R1990) Data Link Encryption.

ANSI X3.106-1983(R1990) Modes of Operation for the Data Encryption Algorithm.

ANSI X3.108-1988 Information Systems—Local Distributed Data Interfaces—Physical Layer Interface to Nonbranching Coaxial Cable Bus.

ANSI X3.139-1987 Information Systems—Fiber Distributed Data interface (FDDI)—Token-Ring Medium Access Control (MAC).

ANSI X3.140-1986 Information Processing Systems—Open Systems Interconnection (OSI)—Connection-Oriented Transport Layer Protocol Specification.

ANSI X3.141-1987 Information Systems—Data Communication Systems and Services—Measurement Methods for User-Oriented Performance Evaluation.

ANSI X3.148 Information systems — Fiber Distributed Data Interface (FDDI) —Token-Ring Physical Layer Protocol.

ANSI X3.153-1987 Information systems—Open Systems Interconnection (OSI)—Basic Connection-Oriented Session Protocol Specification.

ANSI X3.166-1990 Fiber Distributed Data Interface (FDDI) Physical Layer Medium Dependent (PMD).

Data communications standards derived from and identical to the EIA, IEEE, and ISO standards whose numbers they contain

ANSI/EIA 232-D-1986, ANSI/EIA 404-1985 (EIA-404-A), ANSI/EIA 484-1986, ANSI/EIA 530-1987, ANSI/EIA 536-1988, ANSI/EIA 537-1987.

ANSI/IEEE 802.2-1989 (superseded by ISO 8802-2:1990), ANSI/IEEE 802.3-1988 (superseded by ISO 8802-3:1990), ANSI/IEEE 802.4-1990 (superseded by ISO 8802-4:1990), ANSI/IEEE 802.5-1989, ANSI/IEEE 802.7-1989, ANSI/IEEE 1051-1988.

ANSI/ISO 802.3-1988 (superseded by ISO 8802-3:1990), ANSI/ISO 8211-1985.

CCITT RECOMMENDATIONS

CCITT stands for Comité Consultatif International Télégraphique et Téléphonique. The English translation is International Telegraph and Telephone Consultative Committee.

CCITT is one of several organizational entities within the International Telecommunications Union (ITU). Through the work of 17 study groups, it recommends standards for communications equipment interfaces, communications protocols, modem modulation methods, etc. Recommendations fall into several series, each of which uses a numbering scheme characterized by a letter denoting the series, followed by a period and a number. The Recommendations are currently published in the form of "Blue Books," consisting of ten volumes, of which all but two are subdivided into fascicles (bound subvolumes). Fascicle numbers, which use arabic numerals, are qualified by volume numbers, which use Roman numerals (e.g., III.4, for Volume III, Fascicle 4).

This section lists all the I-Series, X-Series, and V-Series Recommendations, with summaries or abstracts for those most frequently encountered by the nonspecialist. The information is correct as of the date of publication of this book.

The following table shows the study groups currently listed by CCITT as active, plus the numbers of the corresponding Blue Book volumes and/or fascicles and the initial letter of the Series associated with their Recommendations.

Study Group	Vol/ Fasc.	Series	Description
-	I.1	-	Minutes and reports of the Plenary Assembly. List of Study Groups and questions under study.
-	I.2	A	Recommendations on the organization and working procedures of CCITT.
-	I.3	B	Terms and definitions. Abbreviations and acronyms. Recommendations on means of expression.
-	I.3	C	General telecommunications statistics.
I	II.4-II.6	F	Definition and operational aspects of telegraph and telematic services (facsimile, telex, videotex, etc.)
II	II.2-II.3	E	Telephone operation and quality of service.
III	II.1	D	General tariff principles.

Study Group	Vol/ Fasc.	Se- ries	Description
IV	IV.1-IV.2	M	General maintenance principles: maintenance of international transmission systems and telephone circuits, international telegraph, phototelegraph and leased circuits, etc.
	IV.3	N	Maintenance of international sound-program and television transmission circuits.
	IV.4	O	Specifications for measuring equipment.
V	IX	K	Protection against interference.
VI	IX	L	Construction, installation, and protection of cable and other elements of outside plant.
VII	VIII.2-VIII.8	X	Data communication networks.
VIII	VII.3-VII.7	T	Terminal equipment for telematic services (facsimile, telex, videotex, etc.).
IX	VII.1	R	Telegraph transmission.
	VII.1	S	Telegraph services terminal equipment.
	VII.2	U	Telegraph switching.
X	X	Z	SDL (Functional Specification and Description Language), CHILL (CCITT High-Level Language) and MML (Man-Machine Language).
XI	VI	Q	Telephone switching and signalling.
XII	V	P	Telephone transmission quality.
XV	III.1-III.5	G	Transmission systems— general characteristics, international analogue carrier systems, transmission media, digital transmission systems, digital networks, etc.
	III.6	H	Line transmission of nontelephone signals. Transmission of sound-program and television signals.
	III.6	J	
XVII	VIII.1	V	Data communication over the telephone network.
XVIII	III.7-III.9	I	Integrated Services Digital Network (ISDN).

CCITT Blue Books are available, in the United States, from the United Nations Bookshop in New York (see p. 166) or, throughout the world, from one of the publishers or resellers in the list starting on p. 176.

I-SERIES: INTEGRATED SERVICES DIGITAL NETWORK (ISDN)

I.110 to I.141: General Structure

I.110 to I.113: Frame of I-Series Recommendations—Terminology

I.110 Preamble and general structure of the I-Series Recommendations.

Introduces the topic of ISDN, noting that it requires a family of CCITT Recommendations to provide principles and guidelines on the ISDN concept, as well as a detailed specification of the user-network and internetwork interfaces.

This Recommendation includes a figure similar to the following. It provides a broad outline of the structure of the I-Series Recommendations and their relationship to other Recommendations.

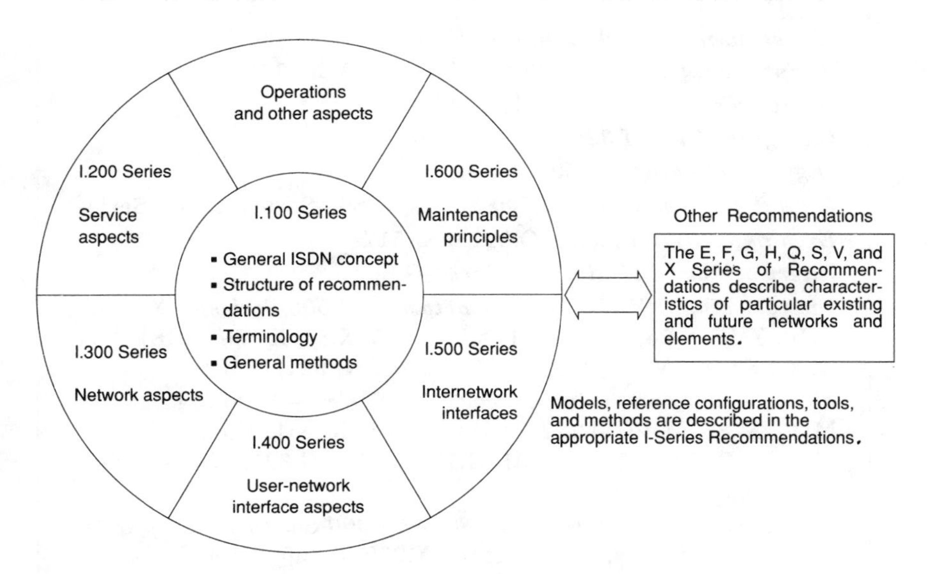

This Recommendation also includes what is essentially a table of contents for the three Blue Books which make up the I-Series Recommendations.

The following pages list all the I-Series Recommendations and, in some cases, include summaries or abstracts.

I.111 Relationship with other Recommendations relevant to ISDNs.

The I-Series Recommendations do not attempt to redefine specific network aspects and ancillary features already covered by

Recommendations in other Series. These may include the following:

- Technical characteristics and performance objectives of component parts (e.g., transmission systems, switching systems, interexchange signalling systems)
- Network synchronization
- Maintenance and operation
- Services
- Tariffs and charging

The references are provided here in their entirety. However, in addition to the I-Series, this book only provides information on the V-Series and X-Series Recommendations (plus some selected Q-Series Recommendations—see the *Index*).

Other Recommendations Containing ISDN-Related Information

Access, user-network (I.400-Series)—Q.920, Q.930
Adaptation, terminal (I.400-Series)—X.30, X.31, V.110, V.120
Bearer services—X.25, X.31, X.300
Charging (I.141, I.326)—D.93, D.200-Series
Digital switching—Q.500-Series
Digital Transmission—G.700-Series, G.800-Series, G.900-Series
Exchange, digital local—Q.511 to Q.517
Interworking: digital hierarchies—G.802
Interworking, ISDN and other networks (I.500 Series)— X.1, X.2, X.10, X.15, X.25, X.30, X.31, X.71, X.75, X.81, X.180, X.181, X.300-Series, V.110, V.120, U.12, U.202, Q.921, Q.931
Management and maintenance (I.600-Series)—M.20 to M.24, M.30, M.36, M.40, M.122, M.125, M.250, M.251, M.550, M.555, M.557, M.770, M.782, G.601, G.700-Series, G.821, G.900-Series, Q.512, Q.542, Q.940
Modelling (I.130-Series, I.140-Series, I.300-Series)— Q.65, Q.71, Q.80, Q.500-Series, Q.700-Series, X.200, X.300, Z.100-Series
Numbering (I.330-Series)—E.163, E.164, E.165, E.166, E.167, F.69, X.121, X.122, X.200, Q.921, Q.931, Q.932, T.90
Parameter exchange (I.515)—V.32, V.100, V.110, V.120, G.725, X.21, X.21 *bis*, X.25, X.30, X.31, Q.931, Q.932, Q.764
Performance (I.350-Series)—G.100-Series, G.821, G.822, G.823, G.824, P.56, P.66, P.84
Protection—K.20, K.22, K.23
Routing (I.335)—E.164, E.170, E.171, E.172, E.502, G.801, X.110, Q.600-Series, Q.700-Series, Q.930, Q.931

Signalling, user-network (I.440-Series, I.450-Series)— Q.920 to Q.940
Signalling, inter-exchange—Q.701 to Q.714, Q.761 to Q.766, Q.771 to Q.774
Speech encoding—G.711, G.721, G.722, G.723, G.725
Supplementary services (I.250-Series)—Q.932, Q.71 to Q.99, X.2
Switching—Q.500-Series
Telephony, transmission quality—G.100-Series
Teleservices (I.240-Series)—E-Series, F-Series, X-Series, T-Series, U.201, G.711, G.722
Terminals (I.470)—T.90, E.330, E.331, P.31, V.110, V.120, V.230
Tones and announcements (I.530)—E.184, V.25
Transmission—G.700-Series, G.800-Series, G.900-Series
Vocabulary (I.112, I.113)—G.701

I.112 Vocabulary of terms for ISDNs.

This Recommendation provides definitions of 99 terms relating to non-broadband ISDN.

I.113 Vocabulary of terms for broadband aspects of ISDNs.

This Recommendation provides definitions of 68 terms relating to broadband ISDN.

I.120 to I.122: Description of ISDNs

I.120 Integrated Services Digital Networks.

Summarizes the principles and the evolution of ISDNs. (For an ISDN summary, see *ISDN* in *Sec. 1.*)

I.121 Broadband aspects of ISDN.

Provides guidelines for the development of further Recommendations for Broadband ISDN (B-ISDN), in the period 1989 to 1992, by Study Group XVIII.

I.122 Framework for providing additional packet-mode bearer services.

Defines 16 special terms and provides recommendations on service aspects, a user-network interface protocol reference model, interworking requirements, the support of OSI connection-oriented network layer service, and applications. Services described include frame relaying 1 and 2, frame switching, and X.25-based additional packet mode.

I.130: General Modelling Methods

I.130 Method for the characterization of telecommunication services supported by an ISDN and network capabilities of an ISDN.

Provides a means to characterize telecommunication services and to define the network capabilities of an ISDN needed to support the identified services. The Recommendation's objectives are:

- To give a common framework and tools for describing services
- To show how, starting from the service definition, one can define protocols and network resources for providing such services
- To make reference to whatever other Recommendations are relevant to the above two points

I.140 to I.141: Telecommunication Network and Service Attributes

I.140 Attribute technique for the characterization of telecommunication services supported by an ISDN and network capabilities of an ISDN.

Attributes may be generic or may relate to services (bearer services, teleservices, supplementary services and charging) or to the network (connection types and connection elements).

The attribute technique described in this Recommendation is used to describe objects in a simple, structured manner and to highlight their important aspects.

Annex A to the Recommendation provides a number of attribute names, defines some of them, and defers definition of others to a later date.

I.141 ISDN network charging capabilities attributes.

This very brief Recommendation covers the method for identifying network charging capabilites and provides a list of candidate attributes.

I.200 to I.257: Service Capabilities

I.200 Guidance to the I.200-Series of Recommendations.

States the objectives of the I.200-Series and summarizes its structure and content.

The objectives given are:

- To achieve a systematic structure which is open to future enhancement
- To help readers by separating the Recommendations covering general service concepts and definitions from those covering detailed characteristics of individual services
- To use, systematically, the service description method given in Recommendation I.130

I.210: General Aspects of Services in ISDN

I.210 Principles of telecommunication services supported by an ISDN and the means to describe them.

Provides classification of the services described by Recommendation I.120, the means for their description based on the method defined in Recommendation I.130, and a basis for the definition of the network capabilities required by an ISDN.

I.220 to I.221: Common Aspects of Services in the ISDN

I.220 Common dynamic description of basic telecommunication services.

From an end-user perspective, and considering the network as a single entity, shows the flow of events and states within the service in a time-sequenced format and identifies all possible relevant actions.

At the time of publication of the Blue Book (1988), only circuit-mode services were described. For packet-mode service descriptions, a revised version of this Recommendation must be ordered.

I.221 Common specific characteristics of services.

Identifies and describes those characteristics which are a common feature of each of the individual services and which help to form a relationship between services.

I.230 to I.232: Bearer Services Supported by an ISDN

I.230 Definition of bearer service categories.

Identifies and describes the following circuit-mode categories, the first three of which are considered essential:

- 64 kbit/s unrestricted, 8 kHz structured

- 64 kbit/s, 8 kHz structured, usable for speech information transfer
- 64 kbit/s, 8 kHz structured, usable for 3.1-kHz audio information transfer
- Alternate speech/64 kbit/s unrestricted, 8 kHz structured
- 2 × 64 kbit/s unrestricted, 8 kHz structured
- 384 kbit/s unrestricted, 8 kHz structured
- 1536 kbit/s unrestricted, 8 kHz structured
- 1920 kbit/s unrestricted, 8 kHz structured

and the following packet-mode categories:

- Virtual call and permanent virtual circuit
- Connectionless (Note)
- User signalling (Note)

Note—The Recommendation does not include descriptions of these two categories, which are for further study.

I.231 Circuit-mode bearer service categories.

Provides a detailed treatment of the eight circuit-mode categories defined in Recommendation I.230.

I.232 Packet-mode bearer service categories.

Provides a detailed treatment of the virtual call and permanent virtual circuit packet-mode category defined in Recommendation I.230.

I.240 to I.241: Teleservices Supported by an ISDN

I.240 Definition of teleservices.

Provides definitions of the following teleservices: Mixed mode, Telefax 4, Telephony, Teletex, Telex, and Videotex.

I.241 Teleservices supported by an ISDN.

Provides service descriptions for the teleservices defined in Recommendation I.240.

I.250 to I.257: Supplementary Services in ISDN

I.250 Definition of supplementary services.

Identifies and describes the services listed here under Recommendations I.251 to I.257. Those followed, here, by an asterisk (*) are not described.

I.251 Number identification supplementary services.

- Direct-Dialling-In (DDI)
- Multiple Subscriber Number (MSN)
- Calling Line Identification Presentation (CLIP)
- Calling Line Identification Restriction (CLIR)
- Connected Line Identification Presentation (COLP)
- Connected Line Identification Restriction (COLR)
- Malicious Call Identification (MCI) *
- Sub-addressing (SUB) *

I.252 Call offering supplementary services.

- Call Transfer (CT)
- Call Forwarding Busy (CFB)
- Call Forwarding No Reply (CFNR)
- Call Forwarding Unconditional (CFU)
- Call Deflection (CD) *
- Line Hunting (LH)

I.253 Call completion supplementary services.

- Call Waiting (CW)
- Call Hold (HOLD)
- Completion of Calls to Busy Subscribers (CCBS) *

I.254 Multiparty supplementary services.

- Conference Calling (CONF)
- Three Party Service (3PTY)

I.255 Community of interest supplementary services.

- Closed User Group (CUG)
- Private Numbering Plan (PNP) *

I.256 Charging supplementary services.

- Credit Card Calling (CRED) *
- Advice of Charge (AOC)
- Reverse Charging (REV) *

I.257 Additional information transfer.

- User-to-User Signalling (UUS)

I.310 to I.352: Overall Network Aspects and Functions

I.310 ISDN—Network functional principles.

Explains ISDN capabilities, including terminal, network, and specialized service center aspects. The word *functional* is used in the sense of an implementation-independent approach.

I.320 to I.326: Reference Models

I.320 ISDN Protocol Reference Model.

States the objective of the Protocol Reference Model (ISDN PRM), which is to model the interconnection and exchange of information (both user and control information) to, through, or inside an ISDN. Compares and contrasts it with the OSI Reference Model (OSI RM).

Introduces the concept of planes, specifically user planes (U-plane) and control planes (C-plane). Further divides C-planes into local (LC-plane) and global (GC-plane).

I.324 ISDN network architecture.

Offers a common functional understanding of the CCITT studies on the general architecture of an ISDN. Deals with reference configurations, which are defined in terms of functional groupings and reference points. See also *Recommendation I.411.*

I.325 Reference configurations for ISDN connection types.

Provides topological descriptions of the detailed network capabilities of the IDSN, as described by the connection types in *Recommendation I.340*. Reference configurations are given, as appropriate.

I.326 Reference configuration for relative network resource requirements.

Provides some initial ideas on the evaluation of the relative network resource requirements associated with the provision of ISDN telecomunication services to subscribers as defined in the I.200-Series.

The discussion (for further study) is based on the following three network resource components:

- Switching capability
- Transmission capability
- Signalling capability

I.330 to I.335: Numbering, Addressing, and Routing

I.330 ISDN numbering and addressing principles.

Deals with the addressing of reference points located at customer premises, the addressing of other functions, and communication with terminals.

I.331 Numbering plan for the ISDN area.

Refers to Recommendation E.164.

I.332 Numbering principles for interworking between ISDNs and dedicated networks with different numbering plans.

Proposes long- and short-term approaches to the resolution of the incompatibilities which exist between ISDN (up to 15 digits) and other numbering plans.

I.333 Terminal selection in ISDN.

Describes procedures for terminal selection in both point-to-point and multipoint operations. *Terminal selection* is defined as the procedures carried out between a terminating ISDN exchange and ISDN terminal equipment situated behind an ISDN interface leading to customer premises. The object of the procedures is to have a terminal response equivalent to an answer or a rejection. Deals with existing terminal types (TE2s), connected via terminal adapters (TAs), and ISDN terminals (TE1s). See definitions in *Recommendation I.411.*

I.334 Principles relating ISDN numbers/subaddresses to the OSI reference model network layer address.

Clarifies the concepts and terminology which relate ISDN numbers and subaddresses to one another and to OSI Reference Model Network layer addresses.

I.335 ISDN routing principles.

Lays down the basic routing principles defining the relationship between ISDN telecommunication services, as described in the I.200-Series of Recommendations, and ISDN network capabilities, as described in the I.300-Series of Recommendations.

I.340: Connection Types

I.340 ISDN connection types.

Identifies and defines the connection types which are a description (using the attribute method of *Recommendation I.140*) of the lower layer functions (described in *Recommendation I.310*) of the ISDN network needed to support the basic services.

I.350 to I.352: Performance Objectives

I.350 General aspects of Quality of Service and Network Performance in digital networks, including ISDN.

This Recommendation:

- Provides descriptions of Quality of Service (QOS) and Network Performance (NP)
- Illustrates how the QOS and NP concepts are applied in digital networks, including ISDNs
- Describes the features of and relationships between these concepts
- Indicates and classifies performance concerns for which parameters may be needed
- Identifies generic performance parameters

I.351 Recommendations in other series concerning network performance objectives that apply at reference point T of an ISDN.

Refers the reader to Recommendation G.821—Error performance of an international digital connection forming part of an integrated services digital network, and Recommendation G.822—Controlled slip rate objectives on an international digital connection.

See *Recommendation I.411* for a definition of *reference point T.*

I.352 Network performance objectives for connection processing delays in an ISDN.

Uses the reference model provided in Recommendation I.340 to provide a baseline reference configuration, with values determined according to Recommendation Q.709.

I.410 to I.470: ISDN User-Network Interfaces

I.410 to I.412: ISDN User-Network Interfaces.

I.410 General aspects and principles relating to Recommendations on ISDN user-network interfaces.

Provides the following examples of user-network interfaces and discusses the related interface objectives, characteristics, and capabilities.

Access to an ISDN of:

- A single ISDN terminal
- A multiple ISDN terminal installation

- Multiservice PBXs or local area networks or, more generally, of private networks
- Specialized storage and information processing centers
- Dedicated service networks
- Other multiple services networks, including ISDNs

I.411 ISDN user-network interfaces—Reference configurations. Restates the definitions of reference configurations, functional groups, and reference points given in *Recommendation I.324* and, among other things, defines the important functional groups and reference points illustrated in the alternative configurations *a* and *b* in the following diagram.

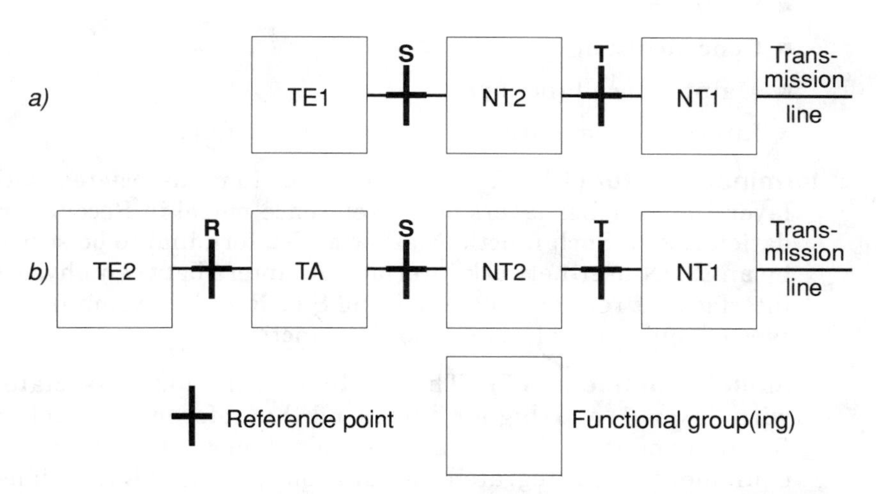

The reference points are implicitly defined with respect to the functional groups on either side. Where the NT2 function is incorporated in a TE1 or a TA, reference points S and T are coincident (S/T reference point). The functional group definitions follow.

network termination 1 (NT1) This includes functions equivalent to layer 1 of the OSI Reference Model and associated with the network's proper physical and electromagnetic termination. NT1 functions are:

- Line transmission termination
- Layer 1 line maintenance functions and performance monitoring
- Timing
- Power transfer

- Layer 1 multiplexing
- Interface termination, including multidrop termination employing layer 1 contention resolution

network termination 2 (NT2) This includes functions equivalent to layer 1 and higher layers of the reference model in Recommendation X.200. Examples of equipment or combinations of equipment that provide NT2 functions are PABXs, local area networks, and terminal controllers. NT2 functions include:

- Layer 2 and layer 3 protocol handling
- Layer 2 and layer 3 multiplexing
- Switching
- Concentration
- Maintenance functions
- Interface termination and other layer 1 functions

terminal adapter (TA) This includes functions associated with layer 1 and higher layers of the reference model in Recommendation X.200. Such functions allow a TE2 terminal to be served by an ISDN user-network interface. Adapters between physical interfaces at reference points R and S or R and T exemplify the type of equipment that provides TA functions.

terminal equipment (TE) This includes functions associated with layer 1 and higher layers of the reference model in Recommendation X.200. Digital telephones, data terminal equipment, and integrated workstations provide TE functions, which are:

- Protocol handling
- Maintenance functions
- Interface functions
- Connection functions to other equipment

terminal equipment type 1 (TE1) This includes functions belonging to the TE functional group, plus an interface complying with the ISDN user-network interface Recommendations.

terminal equipment type 2 (TE2) This includes functions belonging to the TE functional group, but with an interface complying with non-ISDN CCITT Recommendations (e.g., X-

Series) or one that is not included in CCITT Recommendations.

I.412 ISDN user-network interfaces—Interface structures and access capabilities.

Defines limited sets of both channel types and of interface structures for ISDN user-network physical interfaces. (B-, D-, and H-channels are defined in *Sec. 1.*)

I.420 to I.421: Application of I-Series Recommendations to ISDN User-Network Interfaces

I.420 Basic user-network interface.

This Recommendation simply states that the basic user-network interface structure is defined in Recommendation I.412 and that the detailed specifications are in Recommendations I.430 (layer 1), I.440 and I.441 (layer 2), and I.450, I.451, and I.452 (layer 3). (See *Sec. 1* for a definition of the Basic Rate Interface (BRI).)

I.421 Primary rate user-network interface.

This Recommendation simply states that the primary user-network interface structures are defined in Recommendation I.412 and that the detailed specifications are in Recommendations I.431 (layer 1), I.440 and I.441 (layer 2), and I.450, I.451, and I.452 (layer 3).

(See Sec. 1 for a definition of the Primary Rate Interface (PRI).)

I.430 to I.431: ISDN User-Network Interfaces: Layer 1 Recommendations

I.430 Basic user-network interface—Layer 1 specification.

Defines the layer 1 characteristics of the user-network interface to be applied at the S and T reference points for the basic interface structure defined in Recommendation I.412.

Annex E provides definitions of terms that are very specific to this Recommendation.

I.431 Primary rate user-network interface—Layer 1 specification.

Defines the layer 1 characteristics of the user-network interface to be applied at the S and T reference points for the primary rate interface structure defined in Recommendation I.412.

I.440 to I.441: ISDN User-Network Interfaces: Layer 2 Recommendations

I.440 [Q.920] ISDN user-network interface data link layer—general aspects.

I.441 [Q.921] ISDN user-network interface, data link layer specifications.

I.450 to I.452: ISDN User-Network Interfaces: Layer 3 Recommendations

I.450 [Q.930] ISDN user-network interface layer 3—General aspects.

I.451 [Q.931] ISDN user-network interface layer 3 specification for basic call control.

I.452 [Q.932] Generic procedures for the control of ISDN supplementary services.

I.460 to I.465: Multiplexing, Rate Adaption, and Support of Existing Interfaces

I.460 Multiplexing, rate adaption, and support of existing interfaces.

Describes procedures to be used to:

- Adapt the rate of one stream, of rate lower than 64 kbit/s, into a 64 kbit/s B-channel
- Multiplex several streams, of rates lower than 64 kbit/s, into a 64 kbit/s B-channel

I.461 [X.30] Support of X.21, X.21 *bis,* and X.20 *bis* based data terminal equipments (DTEs) by an integrated services digital network (ISDN).

I.462 [X.31] Support of packet mode terminal equipment by an ISDN.

I.463 [V.110] Support of data terminal equipments (DTEs) with V-Series-type interfaces by an integrated services digital network (ISDN).

I.464 Multiplexing, rate adaption, and support of existing interfaces for restricted 64 kbit/s transfer capability.

Defines restricted 64 kbit/s transfer capability (no all-zero octets permitted) and briefly describes its limitations relative to the other I.460-Series Recommendations. Notes the requirement for further study of rate adaption for X.25 DTEs and for DTEs with V-Series interfaces.

I.465 [V.120] Support by an ISDN of data terminal equipment with V-series type interfaces with provision for statistical multiplexing.

I.470: Aspects of ISDN Affecting Terminal Requirements

I.470 Relationship of terminal functions to ISDN.

Provides direction with respect to the functional characteristics required of TE1 and TA devices operating at the basic rate.

I.500 to I.560: Internetwork Interfaces

I.500 General structure of the ISDN interworking Recommendations.

Provides guidance to the organization of the I.500-Series interworking Recommendations and references to other Recommendations that need to be read in conjunction with I.500-Series Recommendations. Organization is at four levels—general, scenario, functional, and protocol.

I.510 Definitions and general principles for ISDN interworking.

I.511 ISDN-to-ISDN layer 1 internetwork interface.

I.515 Parameter exchange for ISDN interworking.

I.520 General arrangements for network interworking between ISDNs.

I.530 Network interworking between an ISDN and a public switched telephone network (PSTN).

I.540 General arrangements for interworking between circuit-switched public data networks (CSPDNs) and integrated services digital networks (ISDNs) for the purposes of data transmission.

I.550 General arrangements for interworking between packet-switched public data networks (PSPDNs) and integrated services digital networks (ISDNs) for the purposes of data transmission.

I.560 Requirements to be met in providing the telex service within the ISDN.

I.601 to I.605: Maintenance Principles

I.601 General maintenance principles of ISDN subscriber access and subscriber installation.

In terms of function groupings and the interconnecting communication parties, outlines the general aspects and principles relating to the reference configurations and general architecture of each kind of subscriber access (basic, primary, multiplexed, higher rate) and associated subscriber installations.

I.602 Application of maintenance principles to ISDN subscriber installations.

I.603 Application of maintenance principles to ISDN basic accesses.

I.604 Application of maintenance principles to ISDN primary rate accesses.

I.605 Application of maintenance principles to static multiplexed ISDN basic accesses.

V-SERIES: DATA COMMUNICATION OVER THE TELEPHONE NETWORK

V.1 to V.7: General

V.1 Equivalence between binary notation symbols and the significant conditions of a two-condition code.

Defines conditions A (space) and Z (mark), their correspondence to the values 0 and 1, respectively, and their representation in amplitude modulation, frequency modulation, phase modulation (reference phase and differential two-phase), and perforated tape.

V.2 Power levels for data transmission over telephone lines.

Defines a maximum power output of subscriber's equipment into the line of 1 mW at any frequency. Also defines maximum power levels for continuous tones (–13 dBm0) and noncontinuous tones (1-min mean of –13 dBm0, maximum instantaneous of 0 dBm0, maximum for 10 Hz bandwidth centered at any frequency of –10 dBm0).

V.3 International Alphabet Number 5. This Recommendation has been superseded by CCITT Recommendation T.50. See *Sec. 5, Interchange Codes*.

V.4 (X.4) General structure of signals of International Alphabet No. 5 code for character-oriented data transmission over public telephone networks.

Covers the use of condition A (space) for 0 and condition Z (mark) for 1 (see *Recommendation V.1*), the use of parity bits, and the use of substitute characters where parity errors are detected. Identical to *Recommendation X.4*.

V.5 Standardization of data signalling rates for synchronous data transmission in the general switched telephone network.

Specifies the use of the following standard rates: 600, 1200, 2400, 4800, and 9600 bit/s.

V.6 Standardization of data signalling rates for synchronous data transmission on leased telephone-type circuits.

Specifies rates in a preferred range—600, 1200, 2400, 4800, 9600, and 14400 bit/s; a supplementary range—3000, 6000, 7200, and 12000 bit/s; and a permitted range—600 x *n* bit/s, where *n* is an integer in the range 1 to 24. Also specifies a maximum deviation from nominal value of ±0.01 percent.

V.7 Definitions of terms concerning data communication over the telephone network.

This Recommendation defines 14 terms that were either new or amended in 1980, 1984, or 1988.

V.10 to V.33: Interfaces and Voice-Band Modems

V.10 [X.26] Electrical characteristics for unbalanced double-current interchange circuits for general use with integrated-circuit equipment in the field of data communications.

Similar to V.11, except that circuits have a common ground. (See description of circuits 102, 102a, and 102b, under *V.24*.)

Corresponds to EIA-423-A. See also *V.11*.

V.11 [X.27] Electrical characteristics for balanced double-current interchange circuits for general use with integrated-circuit equipment in the field of data communications.

Provides for data transfer rates up to 2 Mbit/s. Corresponds to EIA-422-A. V.10 and V.11, together, are intended to overcome the limitations of Recommendation V.28. Improvements in speed and reliability are achieved through the use, in V.10 and V.11, of much lower voltage levels than those specified in V.28, providing for a considerable reduction in the signal distortion caused by cable capacitance. V.11 also eliminates problems relating to ground reference voltage and is recommended for the data and signal-element-timing circuits.

V.13 Simulated carrier control.

Specifies a means by which a DCE's V.24 circuit 105 (request to send—controlled by DTE) can control a remote DCE's circuit 109 (received line signal detector) in cases where it is impossible or impractical for modem carrier to be switched on and off.

V.14 Transmission of start-stop characters over synchronous bearer channels.

Specifies the operation of an asynchronous-to-synchronous converter within or in conjunction with a synchronous DCE.

V.15 Use of acoustic coupling for data transmission.

Specifies the necessary power levels (in dB) for acoustic coupler operation and notes the need for full compliance with the operating characteristics (e.g., *Recommendation V.21*) of the remote DCE.

V.16 Medical analogue data transmission modems.

Defines two modems, one for the simultaneous transmission of the output of up to three ECGs (electrocardiograms) and one, with acoustic coupling, for use with a conventional telephone for the transmission of the output of one ECG. The former modem is defined as having a digital backward channel, supporting frequency-shift keying (FSK) with mark and space frequencies of 390 and 570 Hz, respectively. For the latter modem, a backward digital channel is optional; however, it is recommended that the modem be able, at least, to handle a 389-Hz answering tone.

V.19 Modems for parallel data transmission using telephone signalling frequencies.

Defines a modem which uses the same dual-tone multifrequency signalling as a pushbutton telephone. Transmission speed is limited to about 10 characters per second. See also *DTMF,* in *Sec. 1*.

V.20 Parallel data transmission modems standardized for universal use in the general switched telephone network.

Extends the use of multifrequency signalling beyond the 16-character limit of DTMF, providing for both 64- and 256-character combinations. Transmission speed is limited to 20 characters per second with an intercharacter rest or 40 characters per second with the use of a binary timing channel.

V.21 300-bit/s duplex modem standardized for use in the general switched telephone network.

Specifies full-duplex operation and the use of frequency-shift keying (FSK), with mark and space frequencies of 980 Hz and 1180 Hz, respectively, for the originating modem's transmitted data, and 1650 Hz and 1850 Hz for the answering modem's transmitted data. V.21 is based on a standard developed by Bell, in the United States, for its 103-series modems (which used a different set of frequencies).

V.22 1200-bit/s modem standardized for use in the general switched telephone network and on point-to-point leased telephone-type circuits.

Specifies full-duplex operation and the use of dibit differential phase-shift keying (DPSK). V.22 is based on a standard developed by Bell, in the United States, for its 212-series modems (which used a different set of frequencies and phase shifts). Many auto-answer modems incorporate both the V.21 and V.22 (or, in the United States, both the 212-series and 103-series) capabilities. In such cases, the answering modem adapts to the originating modem's operating characteristics.

V.22 *bis* 2400-bit/s duplex modem using the frequency division technique standardized for use on the general switched telephone network and on point-to-point two-wire leased telephone-type circuits.

Specifies full-duplex operation and the use of quadrature amplitude modulation (QAM), with an originating modem carrier frequency of 1200 Hz and an answering modem carrier frequency of 2400 Hz.

Supports both synchronous and start-stop operation at either 1200 or 2400 bit/s. Incorporates both scrambling and adaptive equalization.

V.22 *bis* operation is commonplace in PC-based modems and is an option with many Hayes and Hayes-compatible modems. Although the Recommendation indicates leased circuit operation, most V.22 *bis* modems operate over dial-up lines.

V.23 600/1200-baud modem standardized for use in the general switched telephone network.

Specifies half-duplex operation and the use of frequency-shift keying (FSK) with 1300 and 1700 Hz mark and space frequencies at 600 bit/s, and 1300 and 2100 Hz mark and space frequencies at 1200 bit/s. A 75-bit/s backward (secondary) channel is specified, with mark and space frequencies of 390 and 450 Hz, respectively. V.23 is based on a standard developed by Bell, in the United States, for its 202-series modems. The 202 series used a different set of frequencies and used the reverse channel only for on/off signalling (at 387 Hz).

V.24 List of definitions for interchange circuits between data terminal equipment (DTE) and data circuit-terminating equipment (DCE).

Circuit definitions follow. The two-character value in parentheses is the equivalent EIA-232-D circuit number. Except for circuit 108/1, those circuits with no EIA-232-D designation are comparatively recent additions.

101 (AA): Protective ground

102 (AB): Signal ground Common return for all other circuits when used in conjunction with Recommendation V.28. DC reference potential for interchange circuits when used in conjunction with Recommendation V.10, V.11, or V.35.

102a: DTE common return Used as a reference potential for unbalanced V.10-type interchange circuit receivers in the DCE.

102b: DCE common return Used as a reference potential for unbalanced V.10-type interchange circuit receivers in the DTE.

102c: Common return Used with single-current (contact closure) interchange circuits as defined by Recommendation V.31.

103 (BA): Transmitted data.

104 (BB): Received data

105 (CA): Request to send Used in half-duplex operation to gain control of the line.

106 (CB): Ready for sending (Clear to send) Used to indicate that the attached DTE may transmit.

107 (CC): Data set ready Used to indicate the DCE's ready status to the DTE.

108/1: Connect data set to line (rarely used)

108/2 (CD): Data terminal ready Used to indicate the DTE's ready status to the DCE.

109 (CF): Received line signal detector Used to indicate the presence of a carrier signal from the remote DCE.

110 (CG): Signal quality detector Used by the DCE to indicate that the incoming signal meets certain standards and is, therefore, unlikely to contain errors.

111 (CH): Data signal rate selector (DTE) Where the DCE can operate at either of two rates, this circuit is used by the DTE to select one of them.

112 (CI): Data signal rate selector (DCE) Where the DCE can operate at either of two rates or rate ranges, this circuit is used by the DCE to indicate the rate to the DTE.

113 (DA): Transmitter signal element timing (DTE) Used by the DTE to provide the DCE with signal element timing information for the data to be transmitted on circuit 103. Use of this circuit is not as common as that of circuit 114 (DB).

114 (DB): Transmitter signal element timing (DCE) Used by the DCE to provide the DTE with signal element timing information for the data to be transmitted on circuit 103 (BA).

115 (DD): Receiver signal element timing Used by the DCE to provide the DTE with signal element timing information of the data being received on circuit 104 (BB).

116/1: Back-up switching in direct mode Used to cause the DCE to switch between normal and standby facilities.

116/2: Back-up switching in authorized mode Used by the DTE to indicate that it is ready to switch from the normal to the standby facility or to cause the DCE to disconnect from the standby facility.

117: Standby indicator Used by the DCE to indicate whether it is conditioned to operate in normal or standby mode.

Backward (secondary) circuits 118 to 123 have the same functions as the corresponding primary circuits (103 to 106, 109, and 110).

118 (SBA): Transmitted backward channel data (secondary transmitted data) Equivalent to circuit 103.

119 (SBB): Received backward channel data (secondary received data) Equivalent to circuit 104.

120 (SCA): Transmit backward channel line signal (secondary request to send) Equivalent to circuit 105.

121 (SCB): Backward channel ready (secondary clear to send) Equivalent to circuit 106.

122 (SCF): Backward channel (secondary) received line signal detector Equivalent to circuit 109.

123: Backward channel signal quality detector Equivalent to circuit 110.

Circuit 124 and circuits 126 to 136 are new. If ever implemented, they will require additional connectors. No detailed explanation is provided in the CCITT text; however, some of them (e.g., circuits 127, 128, 129, and 131) appear to require DCE buffering. Circuit 125 has existed since Recommendation V.24 was first published.

124: Select frequency groups The DTE uses this circuit to select frequency groups from those available in the DCE. The ON condition causes the DCE to use all frequency groups to represent data signals. The OFF condition causes the DCE to use a specified reduced number of frequency groups.

125 (CE): Calling (ring) indicator When ON, this DCE circuit indicates that a calling signal is being received.

126: Select transmit frequency When ON, this DTE circuit selects the higher of two available DCE transmit frequencies. When OFF, it selects the lower of the two frequencies.

127: Select receive frequency When ON, this DTE circuit selects the lower of two available DCE receive frequencies. When OFF, it selects the higher of the two frequencies. Note that this is opposite to the convention used on circuit 126.

128: Receiver signal element timing (DTE source) Used by the DTE to provide the DCE with signal element timing information for the data being received on circuit 104.

129: Request to receive When ON, this DTE circuit causes the DCE to assume the receive mode. When OFF, it causes the DCE to assume the non-receive mode. (Note the careful use of the term *non-receive mode,* which does not imply a transmit mode.)

130: Transmit backward tone When ON, this DTE circuit causes the DCE to transmit a backward channel tone.

131: Received character timing Provides character rather than bit timing.

132: Return to non-data mode Turned ON by the DTE to return the DCE to non-data mode without dropping the line. It is turned OFF once non-data mode has been established.

133: Ready for receiving Used in conjunction with intermediate equipment capable of retaining data. As long as the DTE can receive data, the circuit is turned ON. It is turned OFF during any period when the DTE cannot receive data.

134: Received data present Turned ON by the DCE to indicate that data is being transferred over circuit 104. Turned OFF to indicate the transfer of control information.

136: New signal Turned ON by the DTE to indicate to the DCE that the incoming line signal is about to cease, following which the DCE is to rapidly detect a new line signal.

Circuits 140 to 142 are used with loop test devices as defined in Recommendation V.54.

140: Loopback/maintenance test Turned ON by the DTE to initiate, in the DCE, the maintenance test condition. Turned OFF to terminate that condition.

141: Local loopback Turned ON by the DTE to place the DCE in loop 3 (local analogue loopback, if the DCE is a modem) test mode. Turned OFF to terminate the loop test condition.

142: Test indicator Turned ON by the DCE when it is in a maintenance condition, precluding the transmitting and receiving of data signals. Turned OFF when the DCE is no longer in the maintenance condition.

191: Transmitted voice answer Used for the transmission of DTE-generated analogue signals, typically from an audio-response unit.

192: Received voice answer Used for incoming voice signals, typically generated by a remote audio-response unit.

Circuits 201 to 213 are used with automatic calling equipment conforming to Recommendation V.25. EIA-366-A circuit designations and descriptions, where different, are shown in parentheses.

201 (SGD): Signal ground or common return

202 (CRQ): Call request

203 (DLO): Data line occupied

204 (DSS): Distant station connected (Data set status.)

205 (ACR): Abandon call (and retry)

206 (NB1): Digit signal (2^0) (NB1 digit lead.)

207 (NB2): Digit signal (2^1) (NB2 digit lead.)

208 (NB4): Digit signal (2^2) (NB4 digit lead.)

209 (NB8): Digit signal (2^3) (NB8 digit lead.)

210 (PND): Present next digit

211 (DPR): Digit present

213 (PWI): Power indication

V.25 Automatic answering equipment and/or parallel automatic calling equipment on the general switched telephone network including procedures for disabling of echo control devices for both manually and automatically established calls.

The technique described by this Recommendation (and the U.S. equivalent, EIA-366-A) requires the use of an additional, parallel interface (and, therefore, an additional cable) between the DTE and the DCE. It is no longer used for new equipment.

The circuits used (201 to 213) are described under *V.24*.

V.25 *bis* Automatic calling and/or answering equipment on the general switched telephone network (GSTN) using the 100-Series interchange circuits.

This Recommendation proposes a standard whose general purpose is similar to that of the de facto Hayes protocol. However, it takes into account international considerations, including the limiting of the frequency of unsuccessful attempts to call certain numbers. Also, it does not attempt to define the characteristics of the modem in which it will be incorporated. Call setup involves the use of interchange circuits 106 (ready for sending), 107 (data set ready), and 108/2 (data terminal ready), with messages being exchanged on circuits 103 (transmitted data) and 104 (received data). This contrasts with the Hayes protocol (see *Hayes-compatible modems, Sec. 1*) , which relies entirely on the use of special character sequences on circuit 103 to establish or restore command mode, and assumes that, in command mode, the DTE is prepared to accept responses on circuit 104.

V.26 2400-bit/s modem standardized for use on four-wire leased telephone-type circuits.

Specifies full-duplex operation and the use of dibit differential phase-shift keying (DPSK) with a carrier frequency of 1800 Hz. Allows alternative phase changes (either A (0°, +90°, +180°, and +270°), or B (+45°, +135°, 225°, and 315°), for dibits of 00, 01, 11, and 10, respectively). Defines the same backward channel as *Recommendation V.23.*

V.26 *bis* 2400/1200-bit/s modem standardized for use in the general switched telephone network.

Specifies half-duplex operation using the same modulation technique as in *Recommendation V.26* (alternative B only), and the same backward channel capability as in *Recommendations V.23* and *V.26.* Additionally specifies a fallback rate of 1200 bit/s, using two-phase DPSK (with phase change of 90° for 0 and 270° for 1).

V.26 *ter* 2400-bit/s duplex modem using the echo cancellation technique standardized for use on the general switched telephone network and on point-to-point two-wire leased telephone-type circuits.

Specifies full-duplex (with optional half-duplex) operation using the same modulation technique as in *Recommendation V.26* (alternative A only). Realizes full-duplex capability through echo cancellation, which is defined in *Sec. 1.* Includes a scrambler, an equalizer (compromise or adaptive) and test facilities. Additionally specifies a fallback rate of 1200 bit/s, using 2-phase DPSK (with phase change of 0° for 0 and 180° for 1).

V.27 4800-bit/s modem with manual equalizer standardized for use on leased telephone-type circuits.

Specifies full- or half-duplex operation, using differential eight-phase modulation, with the possibility of a backward channel as specified in *Recommendation V.23.* Uses scrambling and provides a manually adjustable equalizer. Phase changes are in 45° increments from 0° to 315° for tribits of 001, 000, 010, 011, 111, 110, 100, and 101, respectively.

V.27 *bis* 4800/2400-bit/s modem with automatic equalizer standardized for use on leased telephone-type circuits.

Specifies full-duplex (four-wire) or half-duplex (two-wire) operation at 4800 bit/s, using differential eight-phase modulation and scrambling as described in *Recommendation V.27,* and a fallback rate of 2400 bit/s, using differential four-phase modulation as described in *Recommendation V.26* (alternative A). Allows for the possibility of a backward channel as described in *Recommendation V.23*. Includes an automatic adaptive equalizer.

V.27 *ter* 4800/2400-bit/s modem standardized for use in the general switched telephone network.

Specifies the same operation as *Recommendation V.27 bis,* except that it is limited to half-duplex operation (two-wire only).

V.28 Electrical characteristics for unbalanced double-current interchange circuits.

Defines a circuit with a loading between 3000 and 7000 Ω. With this loading, signalling voltages are defined as +15 V (space) and −15 V (mark), with voltages between -5 V and +5 V being ignored (−3 V to +3 V transition region, and a 2-V allowance for noise). It also defines such things as driver power-off output impedance, shunt capacitance, voltage-transition times, and so on.

V.29 9600-bit/s modem standardized for use on point-to-point four-wire leased telephone-type circuits.

Specifies synchronous full- or half-duplex operation at 9600 bit/s, with fallback rates of 7200 and 4800 bit/s, using a combination of phase and amplitude modulation. Includes an automatic adaptive equalizer and a scrambler.

Specifies an optional multiplexer, for combining data rates of 7200, 4800, and 2400 bit/s (with all four arithmetically possible multiplexed configurations at 9600 bit/s, two at 7200 bit/s, and one at 4800 bit/s).

V.31 Electrical characteristics for single-current interchange circuits controlled by contact closure.

V.31 *bis* Electrical characteristics for single-current interchange circuits using optocouplers.

V.32 A family of two-wire, duplex modems operating at data signalling rates up to 9600 bit/s for use on the general switched telephone network and on leased telephone-type circuits.

Specifies full-duplex operation on two-wire circuits using echo cancellation (see *Sec. 1* for definition) for channel separation, with rates selectable from any combination of 9600, 4800, and 2400 bit/s (although 2400 bit/s is for further study). Specifies scrambling and quadrature amplitude modulation (QAM), but provides for two alternative modulation techniques—a mandatory one (for interworking purposes) using 16 carrier states and, optionally, trellis coding with 32 carrier states—at 9600 bit/s.

Provides for optional support of start-stop data streams as specified in *Recommendation V.14.*

V.33 14400-bit/s modem standardized for use on point-to-point four-wire leased telephone-type circuits.

Specifies synchronous full- or half-duplex operation at 14400 bit/s, with a fallback rate of 12000 bit/s, using a combination of phase and amplitude modulation, or optional trellis coding. Includes an automatic adaptive equalizer and a scrambler.

Specifies an optional multiplexer, for combining data rates of 12000, 9600, 7200, 4800, and 2400 bit/s (with all ten arithmetically possible multiplexed configurations at 14400 bit/s and all six at 12000 bit/s).

V.35 to V.37: Wideband Modems

V.35 Data transmission at 48 kbit/s using 60- to 108-kHz group band circuits.

CCITT has declared this Recommendation to be out of date and refers the reader to the newer *Recommendations V.36* and *V.37.* The use of the term *V.35 interface* persists as a reference to the DTE/DCE circuits and the connector used for DTEs and DCEs whose interface specification conforms to *Recommendation V.35.* See also *ISO 2593.*

V.36 Modems for synchronous data transmission using 60- to 108-kHz group band circuits.

This Recommendation specifies operating speeds of 48 kbit/s (with a fallback rate of 40.8 kbit/s), 64 kbit/s, and 72 kbit/s.

Appendix 1 to this Recommendation defines the scrambling process used in this class of modem.

V.37 Synchronous data transmission at a data signalling rate higher than 72 kbit/s using 60- to 108-kHz group band circuits.

Appendix 1 to this Recommendation defines the scrambling process used in this class of modem and, in this context, defines the same four terms as Appendix 1 to *Recommendation V.36.*

V.40 to V.42 *bis*: Error Control and Data Compression

V.40 Error indication with electromechanical equipment.

Based on the limitations of the electromechanical environment, specifies the use of an alarm or error-counting device when, through recognition of incorrect parity, etc., errors are detected.

V.41 Code-independent error-control system.

Defines a scheme, to be used in intermediate equipment (between a DTE and a DCE), using ARQ (automatic request for repeat) techniques in conjunction with fixed block lengths (240, 480, 960, or 3840 bits).

V.42 Error-correcting procedures for DCEs using asynchronous-to-synchronous conversion.

Describes error-correcting protocols for use with V-Series full-duplex DCEs, with conversion of start-stop data from the DTE to synchronous data for transmission (as specified in *Recommendation V.14*). Placement of the error correction function in the DCE relieves the DTE-based data link control function of that responsibility.

V.42 *bis* Data compression procedures for data circuit-terminating equipment (DCE) using error correction procedures.

This Recommendation is not in the 1988 CCITT Blue Book but is obtainable separately. It describes procedures, for use with V-series DCEs, whose principal characteristics are:

- A compression procedure based on an algorithm which encodes strings of characters received from the DTE
- A decoding procedure which recovers the strings of characters from received codewords
- An automatic transparent mode of operation when non-compressible data is detected

An illustration similar to the following is included. It illustrates, in block form, each of the functions comprising a DCE conforming to this Recommendation.

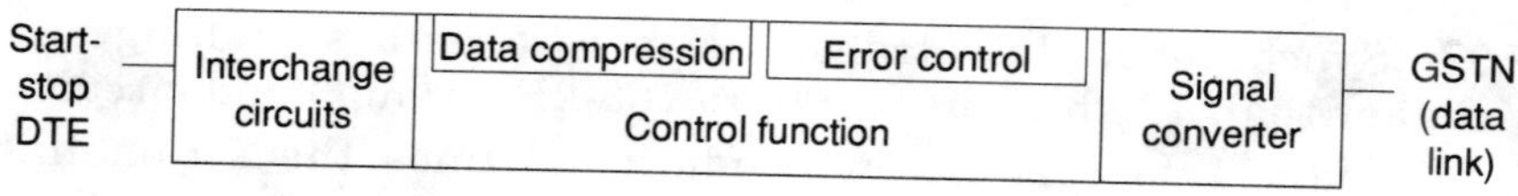

Start-stop DTE — Interchange circuits | Data compression | Error control | Control function | Signal converter — GSTN (data link)

V.50 to V.57: Transmission Quality and Maintenance

V.50 Standard limits for transmission quality of data transmission.

V.51 Organization of the maintenance of international telephone-type circuits used for data transmission.

V.52 Characteristics of distortion and error-rate measuring apparatus for data transmission.

Along with Recommendation V.57, this Recommendation has been replaced by Recommendation O.153.

O.153 deals with methods of measuring error performance for both start-stop and synchronous data circuits at rates from 50 bit/s to 168 kbit/s. Techniques include pseudo-random test patterns, alternating space and mark (0 and 1) states, and plain-language test messages.

V.53 Limits for the maintenance of telephone-type circuits used for data transmission.

V.54 Loop test devices for modems.

Specifies loop testing procedures for the following cases:

- For synchronous mode of operation over point-to-point leased circuit, multipoint, tandem, and general switched telephone network (GSTN) connections
- For start-stop mode of operation over point-to-point leased circuit and GSTN connections

The Recommendation includes a diagram similar to the following, showing the numbering of the four defined loops—two local and two remote. (Loop testing is also discussed under *loopback test* in *Sec. 1.*)

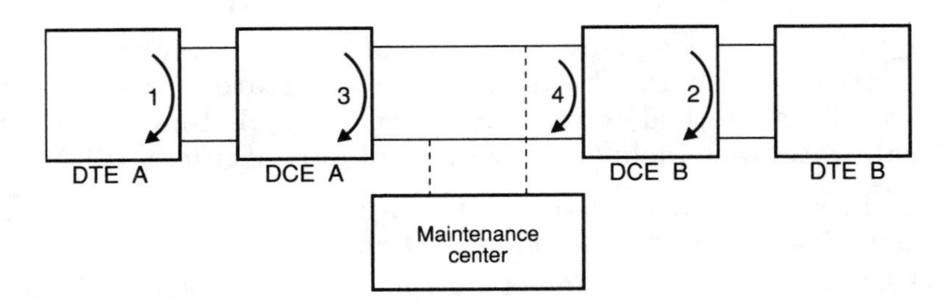

V.55 Specification for an impulse noise measuring instrument for telephone-type circuits.

This Recommendation is identical to Recommendation O.71 and deals with the operating parameters for the measurement of impulse noise.

V.56 Comparative tests of modems for use over telephone-type circuits.

V.57 Comprehensive data test set for high data signalling rates.

Along with Recommendation V.52, this Recommendation has been replaced by Recommendation O.153.

V.100 to V.230: Interworking With Other Networks

V.100 Interconnection between public data networks (PDNs) and the public switched telephone networks (PSTN).

V.110 [I.463] Support of data terminal equipment (DTEs) with V-Series interfaces by an integrated services digital network (ISDN).

Appendix 1 to this Recommendation discusses the subject of in-band parameter exchange (IPE).

V.120 Support by an ISDN of data terminal equipment with V-Series-type interfaces with provision for statistical multiplexing.

V.230 General data communications interface layer 1 specification.

X-SERIES: DATA COMMUNICATION NETWORKS

X.1 to X.32: Services and Facilities, Interfaces

X.1 International user classes of service in public data networks and integrated services digital networks (ISDNs).

By class number, defines classes of service in terms of mode (start-stop and synchronous), code structure (for start-stop), switching method (circuit-switched or packet-switched), and data rate. Also includes services that are specific to ISDN.

X.2 International data transmission services and optional user facilities in public data networks and ISDNs.

Proposes standardization of a large number of user facilities for circuit-switched, packet-switched. and leased-circuit transmission services. The services defined are of the same general character as (but more comprehensive than) those listed in this book under *Recommendations I.251* to *I.257* (p. 92).

X.3 Packet assembly disassembly facility (PAD) in a public data network.

Addresses the need for the interworking of a start-stop DTE with a packet-mode DTE or with another start-stop DTE using a packet-switched data service.

X.4 (V.4) General structure of signals of International Alphabet No. 5 code for character-oriented data transmission over public data networks.

See *Recommendation V.4* (p. 102).

X.10 Categories of access for data terminal equipment (DTE) to public data transmission services.

Defines categories, within 15 lettered groups, for direct and switched access to PDNs (public data networks) and ISDNs (integrated services data networks). Covers the following possibilities:

- Direct connection of the DTE to the public data network or ISDNs
- Switched connection of the DTE to a PDN via an intermediate public network of another type [including PDN, PSTN (public switched telephone network) or ISDN]
- Switched connection of the DTE to an ISDN [including through a terminal adapter (TA)] via an intermediate public network of another type

X.20 Interface between data terminal equipment (DTE) and data circuit-terminating equipment (DCE) for start-stop transmission services on public data networks.

This Recommendation is intended as a replacement for the V.24/V.28/ISO 2110 combination in those cases where transmission will be over a circuit-switched or packet-switched (X.25) public data network. X.20 defines the physical interface. For packet-switched networks, X.28 defines the logical interface protocol. See also *Recommendation X.24*.

X.20 *bis* Use on public data networks of data terminal equipment (DTE) which is designed for interfacing to asynchronous duplex V-Series modems.

Formalizes the application of *Recommendations V.24* and *V.28* as an interim standard for the support of existing data terminal equipment.

X.21 Interface between data terminal equipment (DTE) and data circuit-terminating equipment (DCE) for synchronous operation on public data networks.

X.21 replaces the six control circuits of *Recommendation V.24* with two circuits (“control” from the DTE to the DCE, “indication” from the DCE to the DTE). Control is asserted when the DTE needs to send control information over the transmit circuit. Indication is asserted when the DCE needs to send control and status information over the receive circuit. X.21 also includes information on calling, call clearing, the handling of unsuccessful calls, and the formats for call progress and selection signals.

Electrically, X.21 conforms to *Recommendation V.11*. See also *Recommendation X.24*.

X.21 *bis* Use on public data networks of data terminal equipment (DTE) which is designed for interfacing to synchronous V-Series modems.

This interim Recommendation corresponds to *Recommendations V.24* and *V.28* for lower speeds and to *Recommendation V.35* for speeds of 48 kbit/s and over.

X.22 Multiplex DTE/DCE interface for user classes 3 to 6.

Defines the interface between a DTE and a multiplex DCE, operating at 48 kbit/s and multiplexing a number of *Recommendation X.21* subscriber channels employing synchronous transmission.

X.24 List of definitions for interchange circuits between data terminal equipment (DTE) and data circuit-terminating equipment (DCE) on public data networks.

Defines the interchange circuits used with the interfaces defined in *Recommendations X.20* and *X.21*. For a list of these circuits, see *ISO 4903* on pp. 152 and 154.

X.25 Interface between data terminal equipment (DTE) and data circuit-terminating equipment (DCE) for terminals operating in the packet mode and connected to public data networks by dedicated circuit.

Specifies the behavior of the DCE and a minimum set of requirements for a packet-mode DTE, both with respect to the physical layer and the data link layer (see *Recommendation X.200* for Physical and Data Link layer definitions). Covers the following topics:

- DTE/DCE interface characteristics (physical layer)—X.21, X.21 *bis,* V-Series, X.31
- Link access procedures across the DTE/DCE interface
- HDLC [frame structure, LAP (link access protocol), LAPB (link access protocol balanced), MLP (multilink procedure)]
- Description of the packet layer DTE/DCE interface
- Procedures for virtual circuit services
- Packet formats
- Procedures for optional user facilites (packet layer)

X.26 [V.10] Electrical characteristics for unbalanced double-current interchange circuits for general use with integrated-circuit equipment in the field of data communications. See *Recommendation V.10.*

X.27 [V.11] Electrical characteristics for balanced double-current interchange circuits for general use with integrated-circuit equipment in the field of data communications. See *Recommendation V.11.*

X.28 DTE/DCE interface for a start-stop mode data terminal equipment accessing the packet assembly/disassembly facility (PAD) in a public data network situated in the same country.

Defines procedures for the following actions between a start-stop mode DTE and a PAD:

- the establishment of a national access information path
- character interchange and service initialization
- the exchange of control information
- the exchange of user data

X.29 Procedures for the exchange of control information and user data between a packet assembly/disassembly (PAD) facility and a packet mode DTE or another PAD.

A packet-mode DTE does not need a PAD for access to the network. The relationship defined in X.29 is between a packet-mode DTE and a PAD at the remote site (unlike X.28, which defines a start-stop DTE's relationship with the PAD through which it gains access to the network). It also applies to the relationship between a pair of packet-mode DTEs.

X.30 [I.461] Support of X.21, X.21 *bis,* and X.20 *bis* based data terminal equipment (DTEs) by an integrated services digital network (ISDN).

Defines the functions performed by terminal adapters (TAs) for the three types of DTEs. *Terminal adapter* and related terms are defined under *Recommendation I.411* (pp. 97 to 99).

Appendix I of the Recommendation describes a universal terminal adapter, which can adapt to all rates from 600 bit/s to 64 kbit/s.

X.31 [I.462] Support of packet-mode terminal equipment by an ISDN.

Provides the following definitions:

- The aspects of the packet-mode services provided to the ISDN users in accordance with the bearer services defined in the I-Series Recommendations

- The procedures at the ISDN user-network interface for accessing packet-mode services in alignment with Recommendations I.340, I.431, Q.921, and Q.931
- The terminal adapter's functions for adapting existing X.25 terminals

It does not cover the support of start-stop DTE access by or through an ISDN.

X.32 Interface between data terminal equipment (DTE) and data circuit-terminating equipment (DCE) for terminals operating in the packet mode and accessing a packet-switched public data network through a public switched telephone network or an integrated services digital network or a circuit-switched public data network.

For the classes of service defined in *Recommendation X.1* and the categories of access defined in *Recommendation X.10,* defines the functional and procedural aspects of the DTE/DCE interface for DTEs that access a packet-switched public data network (PSPDN) via public switched networks (PSNs). For the purposes of this Recommendation, a PSN is either a public switched telephone network (PSTN) or an integrated services digital network (ISDN) providing circuit-switched bearer service or a circuit-switched public data network (CSPDN).

X.40 to X.181: Transmission, Signalling, Network Aspects, Maintenance, and Administrative Arrangements

X.40 Standardization of frequency-shift modulated transmission systems for the provision of telegraph and data channels by frequency division of a group.

Defines a method of creating twelve 2400-baud, two 9600-baud, or six 2400-baud plus one 9600-baud derived channels on a single data circuit, using frequency modulation. For 2400-baud channels 1 to 12, the mean frequency ranges from 106 kHz down to 62 kHz (4 kHz separation between channels), with mark state 1 kHz above mean and space state 1 kHz below mean for each channel. For 9600-baud channels 1 and 2, respectively, the mean frequencies are 96 and 72 kHz, with mark state 4 kHz above mean and space state 4 kHz below mean.

X.50 Fundamental parameters of a multiplexing scheme for the international interface between synchronous data networks.

Deals with the interworking of networks that use the following structures for the multiplexing of data streams:

- 8-bit envelope
- Four 8-bit envelope grouping
- 10-bit envelope

The first and last bits of an 8-bit envelope are used for framing and status, respectively, leaving 6 bits for data. This is somewhat inelegant in relation to the customary view of the data stream as a series of octets (and may violate the requirements of some networks). The four 8-bit grouping restores the elegance by accommodating three octets in each group.

The first 2 bits of a 10-bit envelope are used for status and alignment, leaving eight bits for one octet of data.

Both kinds of enveloping involve a bearer channel data rate exceeding the data signalling rate—by 1⁄3 for the 8-bit envelopes and by 1⁄4 for the 10-bit envelopes.

X.50 *bis* Fundamental parameters of a 48-kbit/s user data signalling rate transmission scheme for the international interface between synchronous data networks.

Deals, specifically, with 8-bit envelope structures and defers mixed structures for further study. Specifies a channel data rate of 64 kbit/s (4⁄3 × 48 kbit/s—see *Recommendation X.50*).

X.51 Fundamental parameters of a multiplexing scheme for the international interface between synchronous data networks using 10-bit envelope structure.

Specifies frame length, padding scheme, etc. For standard data rates up to 9600 bit/s (600, 2400, 4800, and 9600), the 10-bit envelope's overhead results in signalling rates of 750, 3000, 6000, and 12000 bit/s, which can be conveniently multiplexed into a 60-kbit/s bit stream. Padding is used to allow transmission on a standard 64 kbit/s international bearer channel.

X.51 *bis* Fundamental parameters of a 48-kbit/s user data signalling rate transmission scheme for the international interface between synchronous data networks using 10-bit envelope structure.

Specifies a transmission scheme consisting of consecutive 10-bit envelopes in which every 16th bit is a padding bit. This achieves the desired bearer channel data rate of 64 kbit/s (5⁄4 × 16⁄15 × 48 kbit/s—see *Recommendation X.50*). Certain aspects of padding and framing are for further study.

X.52 Method of encoding anisochronous signals into a synchronous user bearer.

Deals with the issue of converting from start-stop data streams at 200 or 300 bit/s to synchronous user channels at 600 bit/s (or 2400 bit/s where 600 bit/s channels are not available).

X.53 Numbering of channels on international multiplex links at 64 kbit/s.

Proposes a 4-decimal-digit scheme in which the first digit identifies the multiplexing structure (eighty 8-bit envelope or twenty 8-bit envelope), the second digit identifies the channel rate (600, 2400, 4800, 9600, or 48000 bit/s) and the last two digits indicate the channel position within the frame.

X.54 Allocation of channels on international multiplex links at 64 kbit/s.

Provides a table specifying arbitrary configuration numbers.

X.55 Interface between synchronous data networks using a 6 + 2 envelope structure and single channel per carrier (SCPC) satellite channels.

Defines an interim scheme to interface transmission systems with an 8-bit envelope structure with existing satellite channels operating at 48, 50, or 56 kbit/s.

X.56 Interface between synchronous data networks using an 8 + 2 envelope structure and single channel per carrier (SCPC) satellite channels.

Defines an interim scheme to interface transmission systems with a 10-bit envelope structure with existing satellite channels operating at 48, 50, or 56 kbit/s.

X.57 Method of transmitting a single lower-speed data channel on a 64-kbit/s data stream.

Proposes a scheme in which the 8-bit envelopes related to the low-speed channel are repeated as many times as required to achieve the standard 64-kbit/s bearer channel rate [e.g., 80 times for 600 bit/s which, with enveloping, is increased to 800 bit/s (see *Recommendation X.50*)].

X.58 Fundamental parameters of a multiplexing scheme for the international interface between synchronous nonswitched data networks using no envelope structure.

Proposes a scheme in which a 72-octet frame is transmitted every 10 ms, supporting 24 channels of 2400 bit/s, 12 channels of 4800 bit/s, 6 channels of 9600 bit/s, 3 channels of 19.2 kbit/s, or a combination of such channels.

X.60 Common channel signalling for circuit switched data applications.

States that common channel signalling shall be in accordance with Signalling System No. 7.

X.61 Signalling System No. 7—Data user part.

Defines call control and facility registration- and cancellation-related elements for international common channel signalling for circuit-switched data transmission services.

X.70 Terminal and transit control signalling system for start-stop services on international circuits between anisochronous data networks.

Defines a decentralized terminal and transit control signalling system for start-stop services on international circuits between anisochronous data networks.

X.71 Decentralized terminal and transit control signalling system on international circuits between synchronous data networks.

Counterpart, for synchronous data networks, of *Recommendation X.70.*

X.75 Packet-switched signalling system between public networks providing data transmission services.

This is basically a variant of Recommendation X.25, covering interexchange and internetwork connections. Like X.25, it specifies HDLC-based link access procedures, most notably (and not surprisingly) the multilink procedure (MLP).

X.80 Interworking of interexchange signalling systems for circuit switched data services.

Describes procedures for interworking between Recommendations X.60 and X.71, X.70 and X.71, and X.60 and X.70.

X.81 Interworking between an ISDN circuit-switched and a circuit-switched public data network (CSPDN).

Describes interworking arrangements involving transmission capabilities.

X.82 Detailed arrangements for interworking between CSPDNs and PSPDNs based on Recommendation T.70.

Describes interworking arrangements involving telematic services.

X.92 Hypothetical reference connections for public synchronous data networks.

Provides five hypothetical reference connections to be used for assessing overall customer-to-customer performance objectives, determining some data characteristics requirements of the various items in the connections and setting limits to the impairments such items may introduce. The reference connections apply to circuit-switched services, packet-switched services, and leased-line services in public synchronous data networks.

X.96 Call progress signals in public data networks.

Defines and categorizes call progress signals and indicates their applicability to circuit switching and packet switching (including permanent virtual circuits for progress signals not associated with call establishment or termination).

X.110 International routing principles and routing plan for public data networks.

Provides guidelines for administrations to follow in planning interconnections between public data networks.

X.121 International numbering plan for public data networks.

Annex E to this Recommendation provides definitions of nine related terms.

X.122 Numbering plan interworking between a packet-switched public data network (PSPDN) and an integrated services digital network (ISDN) or public switched telephone network (PSTN) in the short term.

Applies to numbering plan interworking across international boundaries and includes interworking involving more than two networks. Does not include procedures at the human-machine interface or the DTE/DCE interface.

X.130 Call processing delays in public data networks when providing international synchronous circuit-switched data services.

Specifies, in tabular form, objectives for:

- Call connection delay
- Call clearing delay
- Network clear indication delay
- Clear confirmation delay

X.131 Call blocking in public data networks when providing international synchronous circuit-switched data services.

Specifies worst-case values for the probability of nonconnection due to network congestion for connections involving combinations of terrestrial links, satellite links, originating and destination national networks, and an international portion.

X.134 Portion boundaries and packet layer reference events: basis for defining packet-switched performance parameters.

Along with Recommendations X.135 to X.137, this defines performance parameters and values for international packet-switched data communication services.

X.135 Speed of service (delay and throughput) performance values for public data networks when providing international packet-switched services.

Defines protocol-specific speed of service parameters and values associated with each of the three data communications functions—call setup, data transfer, and call clearing. Provides detailed mathematical definitions, based on a network model, of the following terms: *call set-up delay, data packet transfer delay, throughput, steady-state throughput, throughput capacity,* and *clear indication delay*. For the three delay types, *end-to-end, national portion,* and *international portion* variants are considered.

X.136 Accuracy and dependability performance values for public data networks when providing international packet-switched services.

Specifies accuracy and dependability values for national and international virtual connection portions of two types, as depicted in the following table.

Portion Type	Typical Characteristics
National A	Terrestrial connection via an access network section
National B	Connection via an access network section with one satellite circuit or via an access network section and one or more transit network sections
International A	Connection via a direct terrestrial internetwork section
International B	Connection via two satellite ciruits and one transit network section or via one satellite circuit and two or more transit network sections

X.137 Availability performance values for public data networks when providing international packet-switched services.

Sets worst-case limits for availability performance values.

X.140 General quality of service parameters for communication via public data networks.

Deals with effects, observable at network interfaces, and defines them on the basis of protocol-independent events.

X.141 General principles for the detection and correction of errors in public data networks.

Discusses forward error correction (FEC), automatic request for repeat (ARQ) procedures, error detection, and error correction. Includes the FCS (frame check sequence) algorithm and explanatory notes on FCS.

X.150 Principles of maintenance testing for public data networks using data terminal equipment (DTE) and data circuit-terminating equipment (DCE) test loops.

Extends the principles discussed in *Recommendation V.54,* with definitions of nine loops (one of type 1, four of type 3, and two each of types 4 and 2).

X.180 Administrative arrangements for international closed user groups (CUGs).

X.181 Administrative arrangements for the provision of international permanent virtual circuits (PVCs).

X.200 to X.219: Open Systems Interconnection (OSI) Model and Notation, Service Definition

X.200 [ISO 7498:1984] Reference model of Open Systems Interconnection for CCITT applications.

This Recommendation provides a comprehensive discussion of the OSI Model, including its purpose and the functions of each layer. It includes a number of special terms, which are italicized here and whose definitions may be found in the CCITT documentation. Because of copyright restrictions, the definitions are not included here.

Many of the terms used include nonspecific layer references, for which the following notation is used:

- *(N)-layer,* for any specific layer
- *(N+1)-layer,* for the next higher layer
- *(N–1)-layer,* for the next lower layer

The same notation is used for other concepts [e.g., *(N)-entity, (N)-facility, (N)-protocol, (N)-service,* etc. related to these layers].

Where specific layer references are appropriate, the layer names are substituted for *(N)*, *(N+1)* and *(N–1)*. The layer names, by layer number, are:

7. Application
6. Presentation
5. Session
4. Transport
3. Network
2. Data Link
1. Physical

The following is an overview of the layer-by-layer descriptions provided in this Recommendation.

Application layer Provides the sole means of access to the OSI environment for *application-processes*. It serves as a window between correspondent *application-processes* which are using the OSI to exchange meaningful information.

Each *application-process* is represented to its peer by the *application-entity*.

Makes known to the OSI environment all specifiable *application-process* parameters of each OSI environment communications instance.

Application-processes exchange information by means of *application-entities, application-protocols,* and *presentation services* (see *Presentation layer*).

In addition to information transfer, the Application layer provides the following application services:

- For intended communication partners
  - Their identification
  - Determination of their availability
  - Authentication
- Establishment of the authority to communicate
- Agreement on
  - Privacy mechanisms
  - Responsibility for error recovery
  - Procedures for the control of data integrity
- Determination of
  - Cost-allocation methodology
  - Adequacy of resources
  - Acceptable quality of service

- Synchronization of cooperating applications
- Selection of the dialogue discipline, including the initiation and release procedures
- Identification of constraints on data syntax (e.g., character sets, data structure)

Presentation layer Provides for the representation of information that *application-entities* either communicate or refer to in their communication. Two complementary aspects, concerning the general concept of syntax, are covered, namely:

- The representation of data to be transferred between *application-entities*
- The representation of the data structure to which *application-entities* refer in their communication, along with the representations of the actions which may be performed on the data structure

The Presentation layer is not concerned with the semantic content of data, which is the responsibility of *application-entities.* However, it is concerned with the provision, to the *application-entities,* of syntax independence. It allows the *application-entities* to use any syntax by providing transformation between such syntaxes and the common syntax needed for communication between them. The use of a common syntax allows for the standardization of *presentation-protocols.*

The Presentation layer provides *session-services* in the form of *presentation-services* to the *application-entities,* plus:

- Transformation of syntax, which is concerned with code and character set conversions, the modification of the layout of data, and the adaptation of actions on the data structures
- Selection and, if necessary, subsequent modification of syntax

Presentation-services involve the following functions:

- Session establishment requests
- Data transfer
- Negotiation and renegotiation of syntax
- Transformation of syntax, including data transformation, formatting, and special-purpose transformations such as compression
- Session termination requests

Session layer Provides the means necessary for the cooperating *presentation-entities* to organize and synchronize their dialogue and to manage their data exchange. To do this, it provides services to establish a *session-connection* between two *presentation-entities* and to support orderly data exchange interactions, for which the *session-connection* is mapped to and uses a *transport-connection*.

The Session layer provides the following services to the Presentation layer:

- *Session-connection* establishment
- *Session-connection* release
- Normal data exchange
- *Quarantine service*
- Expedited data exchange
- Interaction management
- *Session-connection* synchronization
- Exception reporting

Within the Session layer, there are functions which are performed by *session-entities* in order to provide the *session-services*. They are:

- *Session-connection* to *transport-connection* mapping
- *Session-connection* flow control of the *transport-connection*
- Expedited data transfer
- *Session-connection* recovery
- *Session-connection* release
- Session layer management

Transport layer Through the *transport-service,* provides transparent transfer of data between *session-entities,* relieving them of concern for details of reliability and cost-effectiveness. All Transport layer protocols have end-to-end significance and operate only between OSI end *open systems*.

The Transport layer is not concerned with *routing* and relaying, which are the responsibility of the Network layer.

The Transport layer provides the following services to the Session layer:

- *Transport-connection* establishment
- Data transfer

- *Transport-connection* release

Note that there is not necessarily a one-to-one correspondence between *transport-connections* and *session-connections*. A series of *session-connections* may be established and released during the life of a *transport-connection*. Conversely, the needs of a *session-connection* may be met by a series of established and released *transport-connections*, under the control of either *session-entity*.

Within the Transport layer, there are functions which are performed by *transport-entities* in order to provide the *transport-services*. They are:

- The mapping of *transport-address* to *network-address*
- Establishment and release of *transport-connections*
- End-to-end
  - *Multiplexing* of *transport-connections* over *network-connections*
  - Sequence control on individual connections
  - Error detection and quality of service (QOS) monitoring
  - Error recovery
  - *Segmenting, blocking* and *concatenation*
  - Flow control on individual connections
- Supervisory functions
- *Expedited-transport-service-data-unit* transfer

Network layer Establishes, maintains, and terminates *network-connections* between open systems containing communicating *application-entities* and provides the functional and procedural means to exchange *network-service-data-units* over those *network-connections*. Relieves *transport-entities* of concern for the *routing* and relay considerations associated with the establishment and operation of a given *network-connection* and for the way underlying resources such as *data-link-connections* are used to provide *network-connections*.

The Network layer provides the following services or elements of services to the Transport layer:

- *Network-addresses*
- *Network-connections*
- *Network-connection-endpoint-identifiers*
- *Network-service-data-unit* transfer
- Quality of service (QOS) parameters

- Error notification
- *Sequencing*
- *Flow control*
- *Expedited-network-service-data-unit* transfer (optional)
- *Reset* (optional)
- Release
- Receipt of confirmation (optional)

For the three optional services, the user has to request the service, in response to which the *network-service* provider may either provide the service or indicate that it is not available.

Within the Network layer, a variety of functions provides for a wide range of possibilities, from network-connections supported by point-to-point configurations to network-connections supported by complex combinations of differing subnetworks. For the complex cases, sublayering of the Network layer functions is recommended. The Network layer functions are:

- *Routing* and relaying
- *Network-connections*
- *Network-connection multiplexing*
- *Segmenting* and *blocking*
- Error detection and recovery
- *Sequencing*
- *Flow control*
- Expedited data transfer
- *Reset*
- Service selection
- Network layer management

Data Link layer Establishes, maintains, and releases *data-link-connections* among *network-entities* and to transfer *data-link-service-units*. A *data-link-connection* is built upon one or more *physical-connections*. It detects and possibly corrects errors which may occur in the Physical layer. It also enables the Network layer to control the interconnection of *data-circuits* within the Physical layer.

The Data Link layer provides the following services or elements of services to the Network layer:

- Data-link-connection

- *Data-link-service-data-units*
- *Data-link-connection-endpoint-identifiers*
- *Sequencing*
- Error notification
- *Flow control*
- Quality of service (QOS) parameters (possibly optional)

Within the Data Link layer, the following functions are performed:

- *Data-link-connection* establishment and release
- *Data-link-service-data-unit* mapping
- *Data-link-connection splitting*
- Delimiting and synchronization
- Sequence control
- Error detection and recovery
- *Flow control*
- Identification and parameter exchange
- Control of *data-circuit* interconnection
- Data Link layer management

Physical layer Through mechanical, electrical, functional, and procedural means, activates, maintains, and deactivates *physical-connections* for bit transmission between *data-link-entities*. A *physical-connection* may involve intermediate *open systems,* each relaying bit transmission within the Physical layer. *Physical-entities* are interconnected by means of a physical medium.

The Physical layer provides the following services to the Data Link layer:

- *Physical connection*
- *Physical-service-data-units*
- *Physical-connection-endpoints*
- *Data-circuit* identification
- *Sequencing*
- Fault condition notification
- Quality of service (QOS) parameters

Within the Physical layer, the following functions are performed:

- *Physical-connection* activation and deactivation
- *Physical-service-data-unit* transmission
- Physical layer management

X.208 [ISO/IEC 8824:1990] Specification of Abstract Syntax Notation One (ASN.1).

Abstract Syntax Notation One is used as a semiformal tool to define protocols. This Recommendation deals with the notation itself. *Recommendation X.209* defines a set of encoding rules.

X.209 [ISO/IEC 8825:1990] Specification of basic encoding rules for Abstract Syntax Notation One (ASN.1).

X.210 [ISO/TR 8509:1987] Open Systems Interconnection Layer Service definition conventions.

Provides conventions and terminology to be used for Layer Service definitions of Open Systems Interconnection (OSI).

X.211 [ISO/IEC 10022:1990] Physical Service definition of Open Systems Interconnection for CCITT applications.

Defines the OSI Physical Service in terms of:

- The primitive actions and events of the Service
- The parameters associated with each primitive action and event and the form they take
- The interrelationship between and valid sequences of the actions and events

X.212 [ISO 8886 (draft)] Data Link Service definition for Open Systems Interconnection for CCITT applications.

Defines the OSI Data Link Service in the same terms as shown for the Physical Service in *Recommendation X.211*.

X.213 [ISO 8348:1987-1988] Network Service definition for Open Systems Interconnection for CCITT applications.

Defines the OSI Network Service in the same terms as shown for the Physical Service in *Recommendation X.211*.

X.214 [ISO 8072:1986] Transport Service definition for Open Systems Interconnection for CCITT applications.

Defines the OSI Transport Service in the same terms as shown for the Physical Service in *Recommendation X.211*.

X.215 [ISO 8326:1987] Session Service definition for Open Systems Interconnection for CCITT applications.

Defines the OSI Session Service in the same terms as shown for the Physical Service in *Recommendation X.211.*

X.216 [ISO 8822:1988] Presentation Service definition for open systems interconnection for CCITT applications.

Defines the OSI Presentation Service in the same terms as shown for the Physical Service in *Recommendation X.211.*

X.217 [ISO 8649:1988-1990] Association Control Service definition for Open Systems Interconnection for CCITT applications.

X.218 [ISO 9066-1:1989] Reliable transfer: model and service definition.

X.219 [ISO 9072-1:1989] Remote Operations: model, notation, and service definition.

X.220 to X.290: Open Systems Interconnection (OSI) Protocol Specifications, Conformance Testing

X.220 Use of X.200-Series protocols in CCITT Applications.

Provides a diagram showing the Recommendation numbers applicable at each layer of the OSI reference model.

X.223 [ISO 8878:1987-1990] Use of X.25 to provide the OSI Connection-Mode Network Service for CCITT Applications.

Deals with the following elements of the X.25 Packet Layer Protocol (X.25/PLP):

- The virtual circuit types
- The packet types and fields to be mapped to the primitives and parameters of the OSI Connection-Mode Network Service (OSI CONS)
- The optional user facilities and CCITT-specified DTE facilities

X.224 [ISO/IEC 8073:1988-1989] Transport protocol specification for Open Systems Interconnection for CCITT Applications.

Specifies:

- Five classes of procedures for the connection-oriented transfer of data and control information from one transport-entity to a peer transport-entity:

  Class 0: Simple Class

  Class 1: Basic Error Recovery Class

  Class 2: Multiplexing Class

Class 3: Error Recovery and Multiplexing Class
Class 4: Error Detection and Recovery Class

- The means of negotiating the class of procedures to be used by the transport-entities
- The structure and encoding of the transport-protocol-data-units used for the transfer of data and control information

X.225 [ISO 8327:1987] Session Protocol specification for Open Systems Interconnection for CCITT Applications.

Specifies:

- Procedures for a single protocol for the transfer of data and control information from one session-entity to a peer session-entity
- The means of selecting the functional units to be used by the session entities
- The structure and encoding of the session-protocol-data-units used for the transfer of data and control information

X.226 [ISO 8823:1988] Presentation protocol specification for Open Systems Interconnection for CCITT Applications.

Specifies:

- Procedures for the transfer of data and control information from one presentation-entity to a peer presentation-entity
- The means of selecting, by means of functional units, the procedures to be used by the presentation-entities
- The structure and encoding of the presentation-protocol-data-units used for the transfer of data and control information

X.227 [ISO 8650:1988-1990] Association Control protocol specification for Open Systems Interconnection for CCITT Applications.

Specifies:

- Procedures for the transfer of information relating to the application-association control between application-entities
- The abstract syntax for the representation of the Association Control Service Element (ACSE) application-protocol-data-units (APDUs)

X.228 [ISO 9066-2:1989] Reliable Transfer: Protocol specification.

Defines Reliable Transfer Service Element (RTSE) procedures in terms of :

- The interactions between peer RTSE protocol machines through the use of the Association Control Service Element (ACSE) and the presentation-service
- The interactions between the RTSE protocol machine and its service-user

X.229 [ISO 9072-2:1989] Remote operations: protocol specification.

Defines Remote Operations Service Element (ROSE) procedures in terms of:

- The interactions between peer ROSE protocol machines through the use of Reliable Transfer Service Element (RTSE) services or the presentation-service
- The interactions between the ROSE protocol machine and its service-user

X.244 Procedure for the exchange of protocol identification during virtual call establishment on packet-switched public data networks.

Defines some very simple procedures relating to the call user data field of Call Request and Incoming Call packets and the called user data field of Call Accepted and Call Connected packets.

X.290 [ISO 9646-2 (draft)] OSI conformance testing methodology and framework for protocol Recommendations for CCITT Applications.

Part 1 identifies the phases of the conformance testing process, which are characterized by the following four major roles:

- The specification of abstract test suites for particular OSI protocols
- The derivation of executable test suites and associated testing tools
- The role of a client of a test laboratory, having an implementation of OSI protocols to be tested
- The operation of conformance testing, culminating in the production of a conformance test report which gives the results in terms of the applicable Recommendations and the test suite(s) used

In addition, *Part 1* provides tutorial material, together with the definition of concepts and terms.

Part 2 defines the requirements and provides guidance for the specification of abstract test suites for OSI protocols.

X.300 to X.370: Interworking between Networks, Mobile Data Transmission Systems, Internetwork Management

X.300 General principles for interworking between public networks and between public networks and other networks for the provision of data transmission services.

Does the following:

- Defines principles and detailed arrangements for the interworking of different networks in order to provide a data transmission service
- Specifies, in a general network context, the necessary interaction between elements of user interfaces, inter-exchange signalling systems and other network functions, for the support of data transmission services, telematic services, and the OSI connection-mode network service where appropriate
- Defines the principles for realization of international user facilities and network utilities for data transmission services

X.301 Description of the general arrangements for call control within a subnetwork and between subnetworks for the provision of data transmission services.

X.302 Description of the general arrangements for internal network utilities within a subnetwork and intermediate utilities between subnetworks for the provision of data transmission services.

X.305 Functionalities of subnetworks relating to the support of the OSI connection-mode network service.

X.320 General arrangements for interworking between integrated services digital networks (ISDNs) for the provision of data transmission services.

X.321 General arrangements for interworking between circuit-switched public data networks (CSPDNs) and integrated services digital networks (ISDNs) for the provision of data transmission services.

X.322 General arrangements for interworking between packet-switched public data networks (PSPDNs) and circuit-switched public data networks (CSPDNs) for the provision of data transmission services.

X.323 General arrangements for interworking between packet-switched public data networks (PSPDNs).

X.324 General arrangements for interworking between packet-switched public data networks (PSPDNs) and public mobile systems for the provision of data transmission services.

X.325 General arrangements for interworking between packet-switched public data networks (PSPDNs) and integrated services digital networks (ISDNs) for the provision of data transmission services.

X.326 General arrangements for interworking between packet-switched public data networks (PSPDNs) and common channel signalling network (CCSN).

X.327 General arrangements for interworking between packet-switched public data networks (PSPDNs) and private data networks for the provision of data transmission services.

X.350 General interworking requirements to be met for data transmission in international public mobile satellite systems.

X.351 Special requirements to be met for packet assembly/disassembly facilities (PADs) located at or in association with coast earth stations in the public mobile satellite service.

X.352 Interworking between packet-switched public data networks and public maritime mobile satellite data transmission systems.

X.353 Routing principles for interconnecting public maritime satellite data transmission systems with public data networks.

X.370 Arrangements for the transfer of internetwork management information.

X.400 to X.420: Message-Handling Systems

X.400 [ISO 10021-1:1990] Message Handling System and service overview.

Defines the overall system and service of a Message Handling System (MHS) and provides an MHS overview.

X.402 [ISO 10021-2:1990] Message Handling Systems: Overall architecture.

Provides a statement regarding the purpose of a Message Handling System (MHS). It may be summarized as follows:

- An MHS enables users to exchange messages on a store-and-forward basis
- A message submitted on behalf of an originator is conveyed by the Message Transfer System (MTS) and subsequently delivered to the agents of one or more recipients
- The MTS consists of a number of message transfer agents (MTAs)
- MTAs collectively perform the store-and-forward message transfer function
- Access units (AUs) link the MTS to communication systems of other kinds (e.g., postal systems)
- A user agent (UA) assists the user in the preparation, storage, and display of messages
- A message store (MS) provides the user with optional message storage

The Recommendation also serves as a technical introduction to MHS.

X.403 Message handling systems: Conformance testing.

Provides a "well-chosen" subset of the "virtually infinite" number of tests required to guarantee compliance with a protocol standard.

Also refers the reader to the three following CCITT manuals:

- Conformance Testing Specification Manual for IPMS (P2)
- Conformance Testing Specification Manual for MTS (P1)
- Conformance Testing Specification Manual for RTS

IPMS is Interpersonal Messaging System; RTS is Reliable Transfer Server.

X.407 [ISO 10021-3:1990] Message Handling Systems: Abstract service definition conventions.

Specifies the conventions for abstractly describing a distributed information processing task both macroscopically (abstract model) and microscopically (abstract service). Also specifies principles for concretely realizing abstract models and services, using sample environments designated the *yellow environment* and the *green environment*.

X.408 Message Handling Systems: Encoded information type conversion rules.

Specifies the algorithms used by the Message Handling System (MHS) when converting between different types of encoded information.

X.411 [ISO 10021-4:1990] Message Handling Systems: Message Transfer System: abstract service definition and procedures.

Defines the abstract service provided by the Message Transfer System (MTS) and specifies the procedures to be performed by Message Transfer Agents (MTAs) to ensure the correct distributed operation of the MTS.

X.413 [ISO 10021-5:1990] Message Handling Systems: Message Store: Abstract-service definition.

Defines or describes the following:

- The Message Store (MS) abstract-service
- The general-attribute-types and general-auto-action-types related to the MS
- Procedures for Message Store and port realization

X.419 [ISO 10021-6:1990] Message Handling Systems: Protocol specifications.

Specifies the following access protocols:

- Message Transfer System (MTS) access protocol used between a remote user-agent and the Message Transfer System (MTS) to provide access to the MTS abstract service defined in *Recommendation X.411*
- Message-store access protocol used between a remote user-agent and a message-store (MS) to provide access to the MS abstract service defined in *Recommendation X.413*
- MTS transfer protocol used between message transfer agents (MTAs) to provide the distributed operation of the MTS as defined in *Recommendation X.411*

X.420 [ISO 10021-7:1990] Message Handling Systems: Interpersonal messaging system.

Defines interpersonal messaging, a form of Message Handling tailored for ordinary interpersonal business or private correspondence.

X.500 to X.521: Directory

X.500 [ISO/IEC 9594-1:1990] The Directory—Overview of Concepts, Models, and Services.

This Recommendation provides a definition of the Directory, which may be summarized as follows:

- Provides the directory capabilities required by:
  - OSI applications
  - OSI management processes
  - Other OSI layer entities
  - Telecommunication services
- Capabilities include:
  - "User-friendly naming"
  - "Name-to-address mapping," with dynamic binding between objects and their locations
- Data base characteristics
  - Not a general-purpose data base system
  - May be built on a data base system
  - Much higher frequency of queries than of updates
  - No need for instantaneous global commitment of updates
  - Transient conditions (coexistence of both old and new versions of the same information) are acceptable
  - Enquirer's identity or location does not affect the results of queries (other than through access right limitations or owing to unpropagated updates), rendering the Directory unsuitable for certain telecommunications applications (e.g., some types of routing)

X.501 [ISO/IEC 9594-2:1990] The Directory—Models.

Defines models which provide a conceptual and terminological framework for the other Recommendations which define various aspects of the Directory. Functional and organization models define ways in which the Directory can be distributed, both functionally and administratively. The security model provides the framework within which security features are provided. The information model describes the logical structure of the Directory Information Base (DIB).

X.509 [ISO/IEC 9594-8:1990] The Directory—Authentication framework.

For authentication information, this Recommendation

- Specifies the form in which it is held by The Directory
- Describes how it may be obtained from The Directory
- States the assumptions made about how it is formed and placed in The Directory
- Defines three ways in which applications may use it to perform authentication and describes how other security services may be supported by authentication

X.511 [ISO/IEC 9594-3:1990] The Directory—Abstract Service Definition.

Defines in an abstract way the externally visible service provided by The Directory but does not specify individual implementations or products.

X.518 [ISO/IEC 9594-4:1990] The Directory—Procedures for Distributed Operation.

Specifies the behavior of Directory System Agents (DSAs) participating in the distributed Directory application. The allowed behavior has been designed so as to ensure a consistent service, given a wide distribution of the Directory Information Base (DIB) across many DSAs.

X.519 [ISO/IEC 9594-5:1990] The Directory—Protocol Specifications.

Specifies the Directory Access Protocol (DAP) and the Directory System Protocol (DSP), fulfilling the abstract services specified in *Recommendations X.511* and *X.518.*

X.520 [ISO/IEC 9594-6:1990] The Directory—Selected Attribute Types.

Defines a number of attribute types which may be useful across a range of applications for The Directory.

X.521 [ISO/IEC 9594-7:1990] The Directory—Selected Object Classes.

Defines a number of selected attribute sets and object classes which may be found useful across a range of applications of The Directory.

EIA STANDARDS FOR DATA COMMUNICATIONS

The Electronic Industries Association (EIA) publishes a wide range of electrical, electronic, and related standards. Standards are numbered sequentially, rather than within categories (as with CCITT), hence the fragmented nature of the following (complete) list of data communications standards.

On January 1, 1986, the EIA replaced the prefix "RS" with "EIA-" in its numbering scheme. The style and punctuation of the numbers below are exactly as shown in the EIA's publications.

EIA Data Communication Standards

Note—Only EIA-232-D, 334-A, 334-A-1, 366-A, 404-A, 422-A, 423-A, 449, 485, and 530 are "general" data communications standards. EIA-408, 484, and 491, which deal with numerical control equipment, and EIA-536 and 537, which deal with Group 4 facsimile transmission, are included for completeness.

EIA-232-D Interface between data terminal equipment and data circuit-terminating equipment employing serial binary data interchange.

This is the best-known of all EIA data communications standards. Its scope corresponds to that of CCITT Recommendation V.24 (interchange circuit functional descriptions), plus CCITT Recommendation V.28 (signal characteristics). Additionally, it describes 13 different interface configurations, intended to meet the needs of 15 defined system applications.

This is a revision (January 1987) and superset of the previous standard—EIA-232-C.

The Electronic Industries Association recommends that those ordering EIA-232-D documentation also order their Industrial Electronic Bulletin No. 9, which provides application notes.

For interchange circuit functional descriptions, see *V.24* under *CCITT V-Series Recommendations*.

For signal characteristics, see *V.28* under *CCITT V-Series Recommendations*.

For connector pin assignments, see *ISO 2110,* on p. 147.

EIA-334-A Signal quality at interface between data terminal equipment and synchronous data circuit-terminating equipment for serial data transmission.

EIA-334-A-1 Addendum No. 1 to EIA-334-A and EIA-404-A—Application of signal quality requirements to EIA-449.

EIA-366-A Interface between data terminal equipment and automatic calling equipment for data communication.

As with its CCITT counterpart (Recommendation V.25), equipment is no longer being designed to conform to this obsolete standard, which was last revised in 1979. Instead, automatic calling capability is built into the modem and uses serial signalling via the transmitted data circuit of the EIA-232-D interface. (See *CCITT Recommendation V.25 bis*. See also *801 Automatic Calling Unit,* under *Bell modems* in *Sec. 1.*)

EIA-404-A Standard for start-stop signal quality for nonsynchronous data terminal equipment.

EIA-408 Interface between numerical control equipment and data terminal equipment employing parallel binary data interchange.

EIA-422-A Electrical characteristics of balanced voltage digital interface circuits.

See *CCITT Recommendation V.11.*

EIA-423-A Electrical characteristics of unbalanced voltage digital interface circuits.

See *CCITT Recommendation V.10.*

EIA-449 General-purpose 37- and 9-position interface for data terminal equipment and data circuit-terminating equipment employing serial binary data interchange.

EIA-449-1 Addendum No. 1 to EIA-449.

EIA-484 Electrical and mechanical interface characteristics and line control protocol using communication control characters for serial data link between a direct numerical control system and numerical control equipment employing asynchronous full duplex transmission.

EIA-485 Standard for electrical characteristics of generators and receivers for use in balanced digital multipoint systems.

EIA-491 Interface between a numerical control unit and peripheral equipment employing asynchronous binary data interchange over circuits having EIA-423-A electrical characteristics.

EIA-530 High-speed 25-position interface for data terminal equipment and data circuit-terminating equipment.

EIA-536 General aspects of Group 4 facsimile equipment.

EIA-537 Control procedures for telematic terminals.

Industrial Electronic Bulletins

IEB9 Application notes for Standard EIA-232-D.

This bulletin is dated May 1971. Thus, it actually applies to EIA-232-C (formerly referred to as EIA RS-232C).

IEB11 Fault isolation methods for data communications systems. Dated November 1972.

IEB12 Application notes on interconnection between interface circuits using EIA-449 and EIA-232-D.

Like IEB9, this also predates EIA-232-D. It is dated November 1977.

Interim Standards

EIA/IS-43 Omnibus specification—Local network twisted-pair data communications cable.

This standard, which was adopted for use in the U.S. National Electronic Components Quality Assessment System, relates to LAN twisted-pair cable, using insulated copper conductors in a single PVC or fluorocarbon jacket. It is intended as a building wiring system for Token-Ring and similar networks.

The following supplements to EIA/IS-43 provide detailed specifications for five types of twisted-pair cable used in six different environments:

EIA/IS-43AA Type 1, outdoor

EIA/IS-43AB Type 1, nonplenum

EIA/IS-43AC Type 1, riser

EIA/IS-43AD Type 1, plenum

EIA/IS-43AE Type 2, nonplenum

EIA/IS-43AF Type 2, plenum

EIA/IS-43AG Type 6, office

EIA/IS-43AH Type 8, undercarpet

EIA/IS-43AJ Type 9, plenum

IEEE STANDARDS FOR DATA COMMUNICATIONS

802.2-1989 See *ISO 8802-2*.

802.3-1987 See *ISO 8802-3.*

802.4-1985 See *ISO 8802-4.*

802.5-1989 Standard for Local Area Networks: Token Ring Access Method and Physical Layer Specifications.

Describes the medium access control sublayer and the physical layer functions for a ring-structured network using token passing. Most of IBM's Token-Ring products conform to this standard.

802.7-1989 Recommended Practice for Broadband Local Area Networks.

Specifies minimum acceptable physical, electrical, and mechanical characteristics of a broadband cable medium.

1051-1988 Recommended practice for parameters to characterize digital loop performance.

ISO STANDARDS FOR DATA COMMUNICATIONS

ISO, the International Organization for Standardization, is a federation of national standards bodies. At the beginning of 1991, there were 89 member bodies.

ISO's 172 technical committees and 653 subcommittees, organized and supported by technical secretariats in 33 countries, are responsible for the development or adoption of standards in all fields except electrical and electronic engineering. A very significant and growing field is that of data communications.

Over 140 ISO standards deal with one aspect or another of data communications, ranging from the design of interface connectors to the seemingly esoteric, but necessary, abstract syntax notation used in defining the functioning of the various layers of the OSI reference model.

Many of the ISO standards are derived from CCITT Recommendations; others are derived from standards developed or adopted by member standards bodies, such as the American National Standards Institute (ANSI), the British Standards Institution (BSI), the Deutsches Institut für Normung (DIN), and the Standardiseringskomissionen i Sverige (SIS). The Institute of Electrical and Electronic Engineers (IEEE) and the Electronic Industries Association (EIA) also make major contributions, either directly or through ANSI.

With the exception of the interface connector standards (ISO 2110, ISO 2593, ISO 4902, and ISO 4903) and some cross references, the following list provides only the title description of each standard.

The ISO numbering scheme includes the date of the current version of the standard and, where applicable, of an addendum or corrigendum.

ISO 646:1983 Information processing—ISO 7-bit coded character set for information interchange.

International version of ASCII. The alphabet defined by ISO 646 is also known by its CCITT name—International Alphabet No. 5. See *Sec. 5*.

ISO 1113:1979 Information processing—Representation of the 7-bit coded character set on punched tape.

ISO 1155:1978 Information processing—Use of longitudinal parity to detect errors in information messages.

ISO 1177:1985 Information processing—Character structure for start/stop and synchronous character-oriented transmission.

ISO 1745:1975 Information processing—Basic mode control procedures for data communication systems.

ISO 2022:1986 Information processing—ISO 7-bit and 8-bit coded character sets—Code extension techniques.

ISO 2110:1989 Information technology—Data communication—25-pole DTE/DCE interface connector and contact number assignments.

Defines the cable connectors and the connector pin and socket numbering for DTE-to-DCE interface connections conforming to CCITT Recommendations V.24 and V.28 and to the EIA-232-D standard. The following illustration (not exactly to scale) shows the end-of-cable connectors at the DTE (top) and at the DCE (bottom).

ISO 2110 Connectors

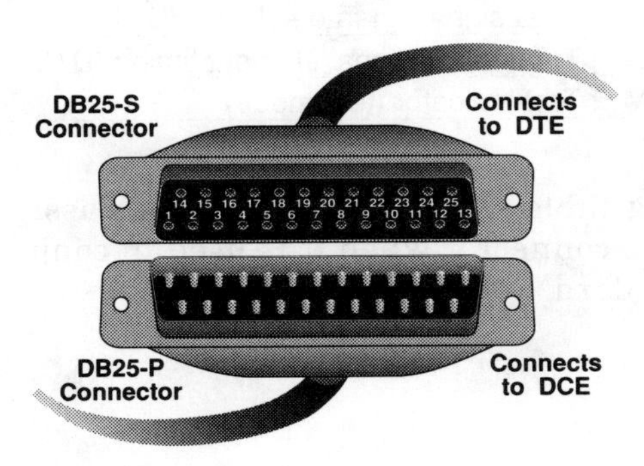

The following table shows the pin numbers assigned to the circuits specified in CCITT Recommendation V.24 (in conjunction with CCITT Recommendation V.28 electrical specifications) and in the EIA-232-D standard (ANSI/EIA 232-D-1986).

ISO 2110 Connector Pin Assignments for CCITT V.24/V.28 Operation

Pin	CCITT V.24	EIA-232-D	Description
1	101	AA	Protective ground (frame)
2	103	BA	Transmitted data
3	104	BB	Received data
4	105	CA	Request to send
5	106	CB	Ready for sending (clear to send)
6	107	CC	Data set ready
7	102	AB	Signal ground or common return
8	109	CF	Data channel received line signal detector
9	—	—	Test voltage, positive *(common practice)*
10	—	—	Test voltage, negative *(common practice)*
11	—	—	(Not assigned)
12	122	SCF	Backward channel received line signal detector
13	121	SCB	Backward channel ready (clear to send)
14	118	SBA	Transmitted backward channel data
15	114	DB	Transmitter signal element timing (DCE)
16	119	SBB	Received backward channel data
17	115	DC	Receiver signal element timing (DCE)
18	141	LL	Local loopback
19	120	SCA	Transmit backward channel line signal (request to send)
20	108/2	CD	Data terminal ready
21	110 140	CG RL	Data signal quality detector or Remote loopback
22	125	CE	Calling (ring) indicator
23	111 112	CH CI	Data signalling rate selector (DTE) or Data signalling rate selector (DCE)
24	113	DA	Transmitter signal element timing (DTE)
25	142	TM	Test indicator (test mode)

The following table shows the pin number assignments associated with this connector when it is used in conjunction with the EIA-530 standard.

ISO 2110 Connector Pin Assignments for EIA-530 Operation

Pin(s)	CCITT V.24	EIA	Description (EIA-530)
1			Shield
2,14	103	BA	Transmitted data
3,16	104	BB	Received data
4,19	105	CA	Request to send
5,13	106	CB	Clear to send
6,22	107	CC	DCE ready (data set ready)
7	102	AB	Signal ground
8,10	109	CF	Received line signal detect (carrier detect)
15,12	114	DB	Transmit signal element timing (DCE source)
17,9	115	DD	Receiver signal element timing (DCE source)
18	141	LL	Local loopback
20,23	108/2	CD	DTE ready (data terminal ready)
21	140	RL	Remote loopback
24,11	113	DA	Transmit signal element timing (DTE source)
25	142	TM	Test mode

ISO 2111:1985 Data communication—Basic mode control procedures—Code-independent data transfer.

ISO 2375:1985 Data processing—Procedure for registration of escape sequences.

ISO maintains an International Register of graphics and control character sets, developed in accordance with ISO 646 and ISO 2022 and assigned escape sequences. ISO 2375 describes the procedures to be followed in registering such escape sequences.

ISO 2593:1984 Data communication—34-pin DTE/DCE interface connector and pin assignments.

This is the connector recommended for connecting DTEs to DCEs conforming to CCITT Recommendation V.35. CCITT Recommendations V.36 and V.37 also allow for its use, on an interim basis.

DTE/DCE interfaces using this connector and conforming logically and electrically to the interface specifications in Recommendation V.35 are referred to as "V.35 interfaces," even where the DCE is not a modem conforming to CCITT Recommendation V.35. The following illustration shows the end-of-cable connectors at the DTE or DCE (top) and at both ends of the cable (bottom).

ISO 2593 Connectors

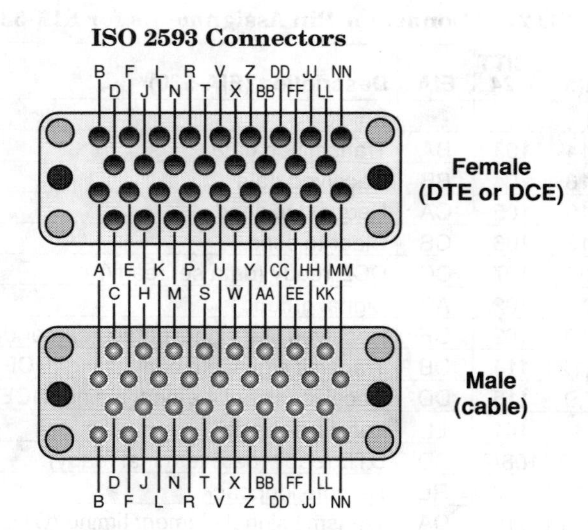

The following table shows the pin numbers assigned to the V.35 subset of the circuits specified in CCITT Recommendation V.24.

ISO 2593 Connector Pin Assignments

Pin(s)	CCITT V.24 circuit	Mnem-onic	Description
A	101	PG	Protective ground (frame)
B	102	SG	Signal ground
C	105	RTS	Request to send
D	106	CTS	Ready for sending (clear to send)
E	107	DSR	Data set ready
F	109	CD	Data channel received line signal detector (carrier detect)
H	108/2	DTR	Data terminal ready
J	125	RI	Calling (ring) indicator
R,T	104	RD	Received data
V,X	115	SCR	Receiver signal element timing
Y,AA	114	SCT	Transmitter signal element timing (DCE)
P,S	103	SD	Transmitted data
U,W	113	SCTE	Transmitter signal element timing (DTE)
Pins K, L, M, N, Z, BB, CC, DD, EE, FF, HH, JJ, KK, LL, MM, and NN are unassigned.			

ISO 2628:1973 Basic mode control procedures—complements.

ISO 2629:1973 Basic mode control procedures—Conversational information message transfer.

ISO 3309:1984 Information processing systems—Data communication—High-level data link control procedures—Frame structure.

ISO 4335:1987 Information processing systems—Data communication—High-level data link control elements of procedures.

ISO 4873:1986 Information processing—ISO 8-bit code for information interchange—Structure and rules for implementation.

ISO 4902:1989 Information technology—Data communication—37-pole DTE/DCE interface connector and contact number assignments.
Defines the cable connectors and the connector pin and socket numbering for DTE-to-DCE interface connections conforming to CCITT Recommendations V.36 and V.37, and the EIA-449 standard. The following illustration (not exactly to scale) shows the end-of-cable connectors at the DTE (top) and at the DCE (bottom).

ISO 4902 Connectors

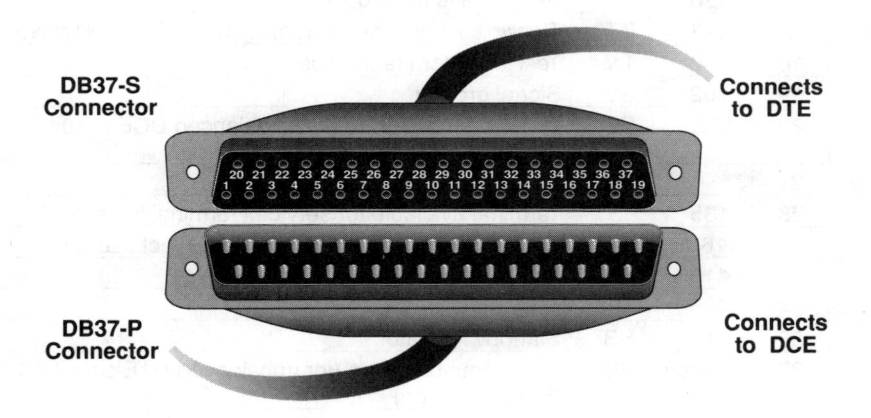

The following table shows the pin numbers assigned to the V.36/V.37/EIA-449 subsets of the circuits specified in CCITT Recommendation V.24. All circuits with two pin numbers are balanced (CCITT Recommendation V.11). Circuits with one pin number are unbalanced (CCITT Recommendation V.10) and use the DTE common return (pin 37) or the DCE common return (pin 20), as appropriate. EIA-449 descriptions, where different from CCITT descriptions, are shown in parentheses.

ISO 4902 Connector Pin Assignments

Pin(s)	CCITT V.24 circuit	EIA-449 Mnem-onic	Description
1			Shield
2	112	SI	Data signal rate selector (DCE)
3			(Unused)
4,22	103	SD	Transmitted data (Send data)
5,23	114	ST	Transmitter signal element timing (DCE) (Send timing)
6,24	104	RD	Received data (Receive data)
7,25	105	RS	Request to send
8,26	115	RT	Receiver signal element timing (Receive timing)
9,27	106	CS	Ready for sending (Clear to send)
10	141	LL	Local loopback
11,29	107	DM	Data set ready (Data mode)
12,30	108/2 or 108/1	TR	Data terminal ready (Terminal ready) or Connect data set to line (rarely used)
13,31	109	RR	Received line signal detector (Receiver ready)
14	140	RL	Loopback/maintenance test (Remote loopback)
15	125	IC	Calling indicator (Incoming call)
16	111 or 126	SR	Data signal rate selector (DTE) (Signal rate selector) or Select transmit frequency
17,35	113	TT	Transmitter signal element timing (DTE) (Terminal timing)
18	142	TM	Test indicator (Test mode)
19	102	SG	Signal ground
20	102b	RC	DCE common return (for unbalanced DCE circuits) (Receive common)
21			(Unused)
28	135	IS	Terminal available for service (Terminal in service)
32	116/1	SS	Backup switching in direct mode (Select standby)
33	110	SQ	Signal quality detector (Signal quality)
34	136	NS	New signal
36	117	SB	Standby indicator
37	102a	SC	DTE common return (for unbalanced DTE circuits) (Send common)

ISO 4903:1989 Information technology—Data communication—15-pole DTE/DCE interface connector and contact number assignments.

Defines the cable connectors and the connector pin and socket numbering for DTE-to-DCE interface connections conforming to CCITT Recommendations X.20 and X.21. The illustration below (not exactly to scale) shows the end-of-cable connectors at the DTE (top) and at the DCE (bottom).

ISO 4903 Connectors

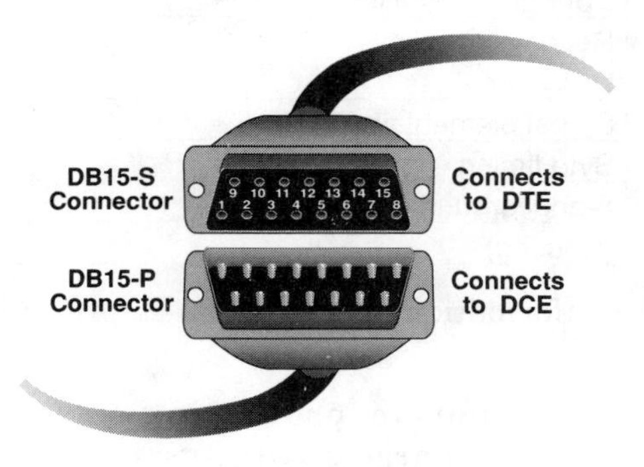

The following table shows the relationship of pin numbers to circuit assignments where the connector is used in conjunction with CCITT Recommendations X.20 and X.26 (V.10).

ISO 4903 Connector Pin Assignments for CCITT X.20 Operation, Assuming the Use of CCITT X.26/V.10 Unbalanced Circuits

Pin	Circuit	Description
1		Protective ground (shield)
2	T	Transmit
3		(Unused)
4	R	Receive
5-7	I	(Unused)
8	G	Signal ground or common return
9	Ga	DTE common return
10		(Unused)
11	Gb	DCE common return
12-15		(Unused)

The only changes for CCITT X.27/V.11 balanced circuit operation are that pins 2 and 9 are used for the transmit circuit (T) and pins 4 and 11 are used for the receive circuit (R).

The following table shows the relationship of pin numbers to circuit assignments where the connector is used in conjunction with CCITT Recommendations X.21 and X.27 (V.11).

ISO 4903 Connector Pin Assignments for CCITT X.21 Operation, Assuming the Use of CCITT X.27/V.11 Balanced Circuits

Pin(s)	Circuit	Description
1		Protective ground (shield)
2, 9	T	Transmit

Pin(s)	Circuit	Description
3,10	C	Control
4,11	R	Receive
5,12	I	Indication
6,13	S	Signal element timing
7,14	B	Byte timing
8	G	Signal ground
15		(Unused)

ISO 5218:1977 Information interchange—Representation of human sexes.

ISO 6093:1985 Information processing—Representation of numerical values in character strings for information interchange.

ISO 6429:1988 Information processing—Control functions for 7-bit and 8-bit character sets.

ISO 6523:1984 Data interchange—Structures for identification of organizations.

ISO 6936:1988 Information processing—Conversion between the two coded character sets of ISO 646 and ISO 6937-2 and the CCITT International Telegraph Alphabet No. 2 (ITA 2).

ISO 6937-1:1983 Information processing—Coded character sets for text communication—Part 1: General introduction.

ISO 6937-2:1983 Information processing—Coded character sets for text communication—Part 2: Latin alphabetic and non-alphabetic graphic characters.

ISO 7350:1984 Text communication—Registration of graphic character subrepertoires.

ISO/TR 7477:1985 Data communication—Arrangements for DTE to DTE physical connection using V.24 and X.24 interchange circuits.

Describes the wiring scheme commonly referred to as a "null modem." *Sec. 1* of this book includes a description of a null modem and a diagram illustrating the wiring scheme for CCITT V.24 (EIA-232-D) circuits.

ISO 7478:1987 Information processing systems—Data communication—Multilink procedures.

ISO 7480:1984 Information processing—Start-stop signal transmission quality at DTE/DCE interfaces.

ISO 7498:1984 [CCITT X.200] Information processing systems—Open Systems Interconnection—Basic reference model.

Technical Corrigendum 1:1988.

Addendum 1:1987 Connectionless-mode transmission.

ISO 7498-2:1989 Information processing systems—Open Systems Interconnection—Basic Reference Model—Part 2: Security Architecture.

ISO 7498-3:1989 Information processing systems—Open Systems Interconnection—Basic Reference Model—Part 3: Naming and addressing.

ISO/IEC 7498-4:1989 Information processing systems—Open Systems Interconnection—Basic Reference Model—Part 4: Management framework.

ISO 7776:1986 Information Processing Systems—Data Communications—High-level data link control procedures—Description of the X.25 LAPB-compatible DTE data link procedures.

Technical Corrigendum 1:1989.

Technical Corrigendum 2:1989.

ISO 7809:1984 Information Processing Systems—Data Communications—High-level data link control procedures—Consolidation of classes of procedures.

Addendum 1:1987.

Addendum 2:1987 Description of optional functions.

ISO 8072:1986 [CCITT X.214] Information processing systems—Open Systems Interconnection—Transport service definition.

Addendum 1:1986 Connectionless-mode transmission.

ISO/IEC 8073:1988 [CCITT X.224] Information processing systems—Open Systems Interconnection—Connection-oriented transport protocol specification.

Technical Corrigenda 1, 2 and 3:1990.

Addendum 1: 1988 Network connection management subprotocol.

Addendum 2: 1989 Class four operation over connectionless network service.

ISO/IEC 8208:1990 Information technology—Data communications—X.25 Packet Layer Protocol for Data Terminal Equipment.

Amendment 1:1990 Alternative Logical Channel Identifier Assignment.

ISO 8326:1987 [CCITT X.215] Information processing systems—Open Systems Interconnection—Basic connection-oriented session service definition.

ISO 8327:1987 [CCITT X.225] Information processing systems—Open Systems Interconnection—Basic connection-oriented session protocol specification.

ISO 8348:1987 [CCITT X.213] Information processing systems—Data communications—Network service definition.

Addendum 1:1987 Connectionless-mode transmission.

Addendum 2:1988 Network layer addressing.

Addendum 3:1988 Additional features of the network service.

ISO 8471:1987 Information processing systems—Data communications—High-level data link control balanced classes of procedures—Data link layer address resolution/negotiation in switched environments.

ISO 8473:1988 Information processing systems—Data communications—Protocol for providing the connectionless-mode network service.

Addendum 3:1989 Provision of the underlying service assumed by ISO 8473 over subnetworks which provide the OSI data link service.

ISO 8480:1987 Information processing—Data communication—DTE/DCE interface back-up control operation using the 25-pole connector.

ISO 8481:1986 Data communication—DTE to DTE physical connection using X.24 interchange circuits with DTE-provided timing.

This describes a CCITT X.24-specific null modem wiring scheme for isochronous communication between synchronous DTEs, similar to the V.24-specific scheme described, under *null modem,* in *Sec. 1.*

ISO 8482:1987 Information processing systems—Data communication—Twisted-pair multipoint interconnections.

ISO/TR 8509:1987 [CCITT X.210] Information processing systems—Open Systems Interconnection—Service conventions.

ISO 8571-1:1988 Information processing systems—Open Systems Interconnection—File Transfer, Access, and Management—Part 1: Introduction.

ISO 8571-2:1988 Information processing systems—Open Systems Interconnection—File Transfer, Access, and Management—Part 2: Virtual Filestore Definition.

ISO 8571-3:1988 Information processing systems—Open Systems Interconnection—File Transfer, Access, and Management—Part 3: File Service Definition.

ISO 8571-4:1988 Information processing systems—Open Systems Interconnection—File Transfer, Access, and Management—Part 4: File Protocol Specification.

ISO/IEC 8571-5:1990 Information processing systems—Open Systems Interconnection—File Transfer, Access and Management—Part 5: Protocol Implementation Conformance Statement Proforma.

ISO 8613-1:1989 Information processing—Text and office systems—Office Document Architecture (ODA) and interchange format—Part 1: Introduction and general principles.

ISO 8613-2:1989 Information processing—Text and office systems—Office Document Architecture (ODA) and interchange format—Part 2: Document structure.

ISO 8613-4:1989 Information processing—Text and office systems—Office Document Architecture (ODA) and interchange format—Part 4: Document profile.

ISO 8613-5:1989 Information processing—Text and office systems—Office Document Architecture (ODA) and interchange format—Part 5: Office Document Interchange Format (ODIF).

ISO 8613-6:1989 Information processing—Text and office systems—Office Document Architecture (ODA) and interchange format— Part 6: Character content architectures.

ISO 8613-7:1989 Information processing—Text and office systems—Office Document Architecture (ODA) and interchange format—Part 7: Raster graphics content architectures.

ISO 8613-8:1989 Information processing—Text and office systems—Office Document Architecture (ODA) and interchange format—Part 8: Geometric graphics content architectures.

ISO 8649:1988 [CCITT X.217] Information processing systems—Open Systems Interconnection—Service Definition for the Association Control Service Element.

Amendment 1:1990 Authentication during association establishment.

ISO 8650:1988 [CCITT X.227] Information processing systems—Open Systems Interconnection—Protocol specification for the Association Control Service Element.

Technical Corrigendum 1:1990

Amendment 1:1990 Authentication during association establishment.

ISO 8802-2:1988 Information processing systems—Local area networks—Part 2: Logical link control.

This describes the functions of the logical link control sublayer, which includes framing, addressing, and error control. It is the basic document used with other standards in the IEEE 802/ISO 8802 series on LANs (except IEEE Standard 802.7-1989). Supersedes *IEEE Standard 802.2-1985*.

ISO/IEC 8802-3:1990 Information processing systems—Local area networks—Part 3: Carrier sense multiple access with collision detection (CSMA/CD) access method and physical layer specifications.

This describes the lower data link layer—medium access control—and the physical layer functions for a bus-structured network using CSMA/CD. This standard was developed in cooperation with Xerox Corporation and is, in fact, a definition of the Ethernet protocols. Supersedes *IEEE Standard 802.3-1987*.

ISO/IEC 8802-4:1990 Information processing systems—Local area networks — Part 4: Token-passing bus access method and physical layer specifications.

This describes the medium access control sublayer and the physical layer functions for a bus-structured network using token passing. Supersedes *IEEE Standard 802.4-1985*.

ISO 8807:1989 Information processing systems—Open Systems Interconnection—LOTOS—A formal description technique based on the temporal ordering of observational behavior.

ISO 8822:1988 [CCITT X.216] Information processing systems—Open Systems Interconnection—Connection-oriented presentation service definition.

ISO 8823:1988 [CCITT X.226] Information processing systems—Open Systems Interconnection—Connection-oriented presentation protocol specification.

ISO/IEC 8824:1990 [CCITT X.208] Information processing systems—Open Systems Interconnection—Specification of Abstract Syntax Notation 1 (ASN.1).

Addendum 1:1990 ASN.1 extensions.

ISO/IEC 8825:1990 [CCITT X.209] Information processing systems—Open Systems Interconnection—Specification of Basic Encoding Rules for Abstract Syntax Notation 1 (ASN.1).

ISO 8831:1989 Information processing systems—Open Systems Interconnection—Job transfer and manipulation concepts and services.

ISO 8832:1989 Information processing systems—Open Systems Interconnection—Specification of the Basic Class Protocol for Job Transfer and Manipulation.

ISO 8859-1:1987 Information processing—8-bit single-byte coded graphic character sets—Part 1: Latin alphabet No. 1.

ISO 8859-2:1987 Information processing—8-bit single-byte coded graphic character sets—Part 2: Latin alphabet No. 2.

ISO 8859-3:1988 Information processing—8-bit single-byte coded graphic character sets—Part 3: Latin alphabet No. 3.

ISO 8859-4:1988 Information processing—8-bit single-byte coded graphic character sets—Part 4: Latin alphabet No. 4.

ISO/IEC 8859-5:1988 Information processing—8-bit single-byte coded graphic character sets—Part 5: Latin/Cyrillic alphabet.

ISO 8859-6:1987 Information processing—8-bit single-byte coded graphic character sets—Part 6: Latin/Arabic alphabet.

ISO 8859-7:1987 Information processing—8-bit single-byte coded graphic character sets—Part 7: Latin/Greek alphabet.

ISO 8859-8:1988 Information processing—8-bit single-byte coded graphic character sets—Part 8: Latin/Hebrew alphabet.

ISO/IEC 8859-9:1989 Information processing—8-bit single-byte coded graphic character sets—Part 9: Latin alphabet No. 5.

ISO 8877:1987 Information processing systems—Interface connector and contact assignments for ISDN basic access interface located at reference points S and T.

ISO 8878:1987 [CCITT X.223] Information processing systems—data communications—Use of X.25 to provide the OSI connection-mode network service.

Technical Corrigenda 1, 2 and 3:1990.

Addendum 1:1990 Priority.

Addendum 2:1990 Use of an X.25 permanent virtual circuit to provide the OSI connection-oriented network service.

ISO/IEC 8880-1:1990 Information technology—Telecommunications and information exchange between systems—Protocol combinations to provide and support the OSI network service—Part 1: General principles.

ISO/IEC 8880-2:1990 Information technology—Telecommunications and information exchange between systems—Protocol combinations to provide and support the OSI network service—Part 2: Provision and support of the connection-mode Network Service.

ISO/IEC 8880-3:1990 Information technology—Telecommunications and information exchange between systems—Protocol combinations to provide and support the OSI network service—Part 3: Provision and support of the connectionless-mode Network Service.

ISO/IEC 8881:1989 Information processing systems—data communications—Use of the X.25 packet-level protocol in local area networks.

ISO 8885:1987 Information processing systems—Data communication—High-level data link control procedures—General-purpose XID frame information field content and format.

Addendum 1:1989 Additional operational parameters for the parameter negotiation data link layer subfield and definition of a multilink parameter negotiation data link layer subfield.

ISO 8886 (draft) [CCITT X.212] Information processing systems—Data communication—Data link service definition for Open Systems Interconnection.

ISO 9036:1987 Information processing—Arabic 7-bit coded character set for information interchange.

ISO 9040:1990 Information technology—Open Systems Interconnection—Virtual Terminal Basic Class Service.

ISO 9041-1:1990 Information technology—Open Systems Interconnection—Virtual Terminal Basic Class Protocol—Part 1: Specification.

ISO 9066-1:1989 [CCITT X.218] Information processing systems—Text communication—Reliable transfer—Part 1: Model and service definition.

ISO 9066-2:1989 [CCITT X.228] Information processing systems—Text communication—Reliable transfer—Part 2: Protocol specification.

ISO 9067:1987 Information processing systems—Data communication—Automatic fault isolation procedures using test loops.

ISO 9070:1990 Information processing—SGML support facilities—Registration procedures for public text owner identifiers.

ISO/IEC 9072-1:1989 [CCITT X.219] Information processing systems—Text communication—Remote Operations—Part 1: Model, notation, and service definition.

ISO/IEC 9072-2:1989 [CCITT X.229] Information processing systems—Text communication—Remote Operations—Part 2: Protocol specification.

ISO 9074:1989 Information processing systems—Open Systems Interconnection—Estelle: A formal description technique based on an extended state transition model.

ISO 9160:1988 Information processing—Data encipherment—Physical layer interoperability requirements.

ISO 9314-1:1989 Information processing systems—Fiber distributed data interface (FDDI)—Part 1: Token Ring Physical Layer Protocol (PHY).

ISO 9314-2:1989 Information processing systems—Fiber distributed data interface (FDDI)—Part 2: Token Ring Medium Access Control (MAC).

ISO/IEC 9314-3:1990 Information processing systems—Fiber distributed data interface (FDDI)—Part 3: Physical Layer Medium Dependent (PMD).

ISO 9542:1988 Information processing systems—Telecommunications and information exchange between systems—End system to intermediate system routing exchange protocol for use in conjunction with the Protocol for providing the connectionless-mode network service (ISO 8473).

ISO 9543:1989 Information processing systems—Information exchange between systems—Synchronous transmission signal quality at DTE/DCE interfaces.

ISO/IEC 9545:1989 Information technology—Open Systems Interconnection—Application Layer structure.

ISO/IEC 9549:1990 Information technology—Galvanic isolation of balanced interchange circuit.

ISO/IEX/TR 9571:1989 Information technology—Open Systems Interconnection—LOTOS description of the session service.

ISO/IEC/TR 9572:1989 Information technology—Open Systems Interconnection—LOTOS description of the session protocol.

ISO/IEC 9574:1989 Information technology—Telecommunications and information exchange between systems—Provision of the OSI connection-mode network service by packet-mode terminal equipment connected to an integrated services digital network (ISDN).

ISO/IEC/TR 9575:1990 Information technology—Telecommunications and information exchange between systems—OSI Routing Framework.

ISO/IEC/TR 9577:1990 Information technology—Telecommunications and information exchange between systems—Protocol identification in the network layer.

ISO/IEC/TR 9578:1990 Information technology—Communication interface connectors used in local area networks.

ISO/IEC 9594-1:1990 [CCITT X.500] Information technology—Open Systems Interconnection—The Directory—Part 1: Overview of concepts, models and services.

ISO/IEC 9594-2:1990 [CCITT X.501] Information technology—Open Systems Interconnection—The Directory—Part 2: Models.

ISO/IEC 9594-3:1990 [CCITT X.511] Information technology—Open Systems Interconnection—The Directory—Part 3: Abstract service definition.

ISO/IEC 9594-4:1990 [CCITT X.518] Information technology—Open Systems Interconnection—The Directory—Part 4: Procedures for distributed operation.

ISO/IEC 9594-5:1990 [CCITT X.519] Information technology—Open Systems Interconnection—The Directory—Part 5: Protocol specifications.

ISO/IEC 9594-6:1990 [CCITT X.520] Information technology—Open Systems Interconnection—The Directory—Part 6: Selected attribute types.

ISO/IEC 9594-7:1990 [CCITT X.521] Information technology—Open Systems Interconnection—The Directory—Part 7: Selected object classes.

ISO/IEC 9594-8:1990 [CCITT X.509] Information technology—Open Systems Interconnection — The Directory—Part 8: Authentication framework.

ISO/IEC 9595:1990 Information technology—Open Systems Interconnection—Common management information service definition.

ISO/IEC 9596:1990 Information technology—Open Systems Interconnection—Common management information protocol specification.

ISO 9646-1 (draft) [CCITT X.290] Information processing systems—Open Systems Interconnection—OSI conformance testing methodology and framework—Part 1: General concepts.

ISO 9646-2 (draft) [CCITT X.290] Information processing systems—Open Systems Interconnection—OSI conformance testing methodology and framework—Part 2: Abstract test suite specification.

ISO/IEC 9804:1990 Information technology—Open Systems Interconnection—Service definition for the Commitment, Concurrency, and Recovery service element.

ISO/IEC 9805:1990 Information technology—Open Systems Interconnection—Protocol specification for the Commitment, Concurrency, and Recovery service element.

ISO/IEC 9834-3:1990 Information technology—Open Systems Interconnection—Procedures for the operation of OSI Registration Authorities—Part 3: Registration of object identifier component values for joint ISO-CCITT use.

ISO/IEC/TR 10000-1:1990 Information technology—Framework and taxonomy of International Standardized Profiles—Part 1: Framework.

ISO/IEC/TR 10000-2:1990 Information technology—Framework and taxonomy of International Standardized Profiles—Part 2: Taxonomy of Profiles.

ISO/IEC 10021-1:1990 [CCITT X.400] Information technology—Text communication— Message-Oriented Text Interchange Systems (MOTIS)—Part 1: System and Service Overview.

ISO/IEC 10021-2:1990 [CCITT X.402] Information technology—Text communication— Message-Oriented Text Interchange Systems (MOTIS)—Part 2: Overall Architecture.

ISO/IEC 10021-3:1990 [CCITT X.407] Information technology—Text communication— Message-Oriented Text Interchange Systems (MOTIS)—Part 3: Abstract Service Definition Conventions.

ISO/IEC 10021-4:1990 [CCITT X.411] Information technology—Text communication—Message-Oriented Text Interchange Systems (MOTIS)—Part 4: Message Transfer System: Abstract Service Definition and Procedures.

ISO/IEC 10021-5:1990 [CCITT X.413] Information technology—Text communication—Message-Oriented Text Interchange Systems (MOTIS)—Part 5: Message Store: Abstract Service Definition.

ISO/IEC 10021-6:1990 [CCITT X.419] Information technology—Text communication—Message-Oriented Text Interchange Systems (MOTIS)—Part 6: Protocol Specifications.

ISO/IEC 10021-7:1990 [CCITT X.420] Information technology—Text communication—Message-Oriented Text Interchange Systems (MOTIS)—Part 7: Interpersonal Messaging System.

ISO/IEC 10022:1990 [CCITT X.211] Information technology—Open Systems Interconnection—Physical Service Definition.

ISO/IEC 10027:1990 Information technology—Information Resource Dictionary System (IRDS) framework.

ISO/IEC/TR 10029:1989 Information technology—Telecommunications and information exchange between systems—Operation of an X.25 interworking unit.

ISO/IEC 10030:1990 Information technology—Telecommunications and information exchange between systems—End System Routing Information Exchange Protocol for use in conjunction with ISO 8878.

ISO/IEC ISP 10607-1:1990 Information technology—International Standardized Profiles AFTnn—File Transfer, Access, and Management—Part 1: Specification of ACSE, Presentation and Session Protocols for use by FTAM.

ISO/IEC ISP 10607-2:1990 Information technology—International Standardized Profiles AFTnn—File Transfer, Access, and Management—Part 2: Definition of document types, constraint sets and syntaxes.

ISO/IEC ISP 10607-3:1990 Information technology—International Standardized Profiles AFTnn—File Transfer, Access, and Management—Part 3: AFT11—Simple File Transfer Service (unstructured).

INTERNATIONAL STANDARDS ORGANIZATIONS

Publications and, in some cases, information on standards may be obtained from the following addresses. Operating hours and catalogue prices are correct at the time of going to press.

CCITT—Comité Consultatif International Télégraphique et Téléphonique (International Telegraph and Telephone Consultative Committee)

See *ITU,* below.

ECMA—European Computer Manufacturers' Association

ECMA
Rue du Rhône 114
CH-1204 Genève
Switzerland
Phone: +41 22 735 36 34
Telex: 22 288 ECMA CH

IEEE—The Institute of Electrical and Electronics Engineers, Inc.

IEEE describes itself as a "transnational" organization, with members in about 130 countries. Its international influence often precedes (by several years) the formal adoption, by ISO, of its proposed standards (e.g., IEEE 488 and the 802 series). For this reason, it is included here, rather than under *National Standards Organizations.*

IEEE Service Center
445 Hoes Lane
P.O. Box 1331
Piscataway, NJ 08855-1331
Phone: (800) 678-4333 (within U.S.)
(908) 981-0060 (international)
Fax: (908) 981-9667 (U.S. and international)
Hours: 8:00 to 16:00, Eastern Time.

Publications

IEEE Publications Catalog: Free.
Payment terms: Check or money order (US$) with order. Credit card telephone orders. Billing is available to IEEE members (subject to credit limit) and to businesses (faxed purchase order, subject to credit approval). Sales tax applies to orders from CA, DC, NJ, and NY.
Shipping is by UPS in the U.S., surface elsewhere. Priority shipping is available at actual cost.

ISO—International Organization for Standardization

ISO
1, rue de Varembé
Case postale 56
CH-1211 Genève 20
Switzerland
Phone: +41 22 749 01 11
Fax: +41 22 733 34 30
Telex: 41 22 05 ISO CH

For U.S. orders, see *ANSI* (p. 175) or one of the U.S.-based publishers starting on p. 177.

ITU (UIT)—International Telecommunications Union

ITU
General Secretariat
Sales and Service
Place des Nations
CH-1211 Genève 20
Switzerland
Phone: +41 22 730 51 11
Fax: +41 22 733 72 56
Telex: 42 10 00 UIT CH

NATIONAL STANDARDS ORGANIZATIONS (Members and Correspondent Members of ISO)

This list includes the United States in its correct alphabetical position. The symbol ♦ indicates that the organization is its country's sales agent for ISO publications.

The symbol ✉ indicates a country with a correspondent member (i.e., not a member body).

Albania—DSMA ♦

Drejtoria e Standardeve dhe e Mjeteve Matësë në Ministrinë e Ekonomisë
Bulevardi: Dëshmorët e Kombit
Tirana
Albania
Phone: +355 42 2 62 55
Telex: 42 95 KOPLAN AB

Algeria—INAPI ♦

Institut algérien de normalisation et de propriété industrielle
5, rue Abou Hamou Moussa
B.P. 1021—Centre de Tri
Alger
Algeria
Phone: +213 2 63 51 80
Fax: +213 2 61 09 71
Telex: 6 64 09 INAPI DZ

Argentina—IRAM

Instituto Argentino de Racionalización de Materiales
Chile 1192
1098 Buenos Aires
Argentina
Phone: +54 1 383 37 51
Fax: +54 1 383 84 63
Telex: 2 60 86 IFLEX AR

Australia—SAA ♦

Standards Australia
P.O. Box 458
North Sydney, NSW 2059
Australia
Phone: +61 2 963 41 11
Fax: +61 2 959 38 96
Telex: 2 65 14 ASTAN AA

Austria—ÖN ♦

Österreichisches Normungsinstitut
Heinestraße 38, Postfach 130
A-1021 Wien
Austria
Phone: +43 1 26 75 35
Fax: +43 1 26 75 52
Telex: 11 59 60 NORM A

Bahrain ✉

Directorate of Standards and Metrology
Ministry of Commerce and Agriculture
P.O. Box 5479
Bahrain
Phone: +973 53 01 00
Fax: +973 53 04 55
Telex: 91 71 TEJARA BN

Bangladesh—BSTI ♦

Bangladesh Standards and Testing Institution
116-A, Tejgaon Industrial Area
Dhaka 1208
Bangladesh
Phone: +880 2 88 14 62

Barbados ✉

Barbados National Standards Institution (BNSI)
"Flodden"
Culloden Road
St. Michael
Barbados
Phone: +500 426 3870
(809) 426 3870 (from U.S.A. or Canada)
Fax: +500 436 1495
(809) 436 1495 (from U.S.A. or Canada)

Belgium—IBN ♦

Institut belge de normalisation
Avenue de la Brabançonne, 29
B-1040 Bruxelles
Belgium
Phone: +32 2 734 92 05
Fax: +32 2 733 42 64
Telex: 2 38 77 BENOR B

Brazil—ABNT ♦

Associação Brasileira de Normas Técnicas
Avenida 13 de Maio, Nº 13-28º andar
Caixa Postal 1680
CEP: 20.003 Rio de Janeiro—RJ
Brazil
Phone: +55 21 210 31 22
Fax: +55 21 532 21 43
Telex: 213 43 33 ABNT BR

Brunei Darussalam ✉

Construction Planning and Research Unit
Ministry of Development
Negara
Brunei Darussalam
Phone: +673 2 24 20 33
Fax: +673 2 24 22 67
Telex: 27 22 MIDEV BU

Bulgaria—BDS ♦

Comité de normalisation, certification et métrologie auprès du Conseil des Ministres
21, rue du 6 Septembre
1000 Sofia
Bulgaria
Phone: +359 2 85 91
Fax: +359 2 80 14 02
Telex: 2 25 70 DKS BG

Canada—SCC ♦

Standards Council of Canada
45 O'Connor Street, Suite 1200
Ottawa, Ontario K1P 6N7
Canada
Phone: (613) 238-3222
Fax: (613) 995-4564
Telex: 053 4403 STANCAN OTT

Chile—INN

Instituto Nacional de Normalización
Matías Cousiño 64, 6º piso
Casilla 955, Correo Central
Santiago
Chile
Phone: +56 2 696 81 44
Fax: +56 2 696 02 47

China, People's Republic of—CSBTS ♦

China State Bureau of Technical Supervision
4, Zhi Chun Road, P.O. Box 8010
Haidian District, Beijing
People's Republic of China
Phone: +86 1 89 49 05
Fax: +86 1 831 26 89
Telex: 22 29 28 SSBTS CN

Colombia—ICONTEC ♦

Instituto Colombiano de Normas Técnicas
Carrera 37, Nº 52-95, Edificio ICONTEC
P.O. Box 14237
Santafé de Bogotá
Colombia
Phone: +57 1 222 05 71
Fax: +57 1 222 14 35
Telex: 4 25 00 ICONT CO

Cuba—NC ♦

Comité Estatal de Normalización
Egido 610, entre Gloria y Apodaca
Zona postal 2
La Habana
Cuba
Phone: +53 7 62 15 03
Fax: +53 7 62 76 57
Telex: 51 22 45 CEN CU

Cyprus—CYS ♦

Cyprus Organization for Standards and Control of Quality
Ministry of Commerce and Industry
Nicosia
Cyprus
Phone: +357 2 30 34 41
Fax: +357 2 36 61 20
Telex: 22 83 MINCOMIN CY

Czechoslovakia—CSN ♦

Federal Office for Standards and Measurements
Václavské námesti 19
113 47 Praha 1
Czechoslovakia
Phone: +42 2 235 21 52
Fax: +42 2 26 57 95
Telex: 12 19 48 FUNM C

Denmark—DS ♦

Dansk Standardiseringsraad
Baunegaardsvej 73
DK-2900 Hellerup
Denmark
Phone: +45 31 77 01 01
Fax: +45 31 77 02 02
Telex: 11 92 03 DS STAND

Egypt—EOS ♦

Egyptian Organization for Standardization and Quality Control
2 Latin America Street
Garden City
Cairo
Egypt
Phone: +20 2 354 97 20
Fax: +20 2 355 78 41
Telex: 9 32 96 EOS UN

Ethiopa—ESA ♦

Ethiopian Authority for Standardization
P.O. Box 2310
Addis Ababa
Ethiopa
Phone: +251 18 51 06
Telex: 2 17 25 ETHSA ET

Finland—SFS ♦

Suomen Standardisoimisliitto SFS
P.O. Box 205
SF-00121 Helsinki
Finland
Phone: +358 0 64 56 01
Fax: +358 0 64 31 47
Telex: 12 23 03 STAND SF

France—AFNOR ♦

Association française de normalisation
Tour Europe
Cédex 7
F-92049 Paris La Défense
France
Phone: +33 1 42 91 55 55
Fax: +33 1 42 91 56 56
Telex: 61 19 74 AFNOR F

Germany—DIN ♦

Deutsche Institut für Normung
Burggrafenstraße 6
Postfach 11 07
D-1000 Berlin 30
Phone: +49 30 26 01-1
Fax: +49 30 26 01-231
Telex: 18 42 73 DIN D

Ghana—GSB ♦

Ghana Standards Board
P.O. Box M-245
Accra
Ghana
Phone: +233 21 66 26 06
Telex: 24 45 MINCOM GH

Greece—ELOT ♦

Hellenic Organization for Standardization
313 Acharnon Street
GR-111 45 Athens
Greece
Phone: +30 1 201 50 25
Fax: +30 1 202 07 76
Telex: 21 96 70 ELOT GR

Hong Kong ✉

Industry Department
Hong Kong Government
14/F Ocean Centre
5 Canton Road
Kowloon
Hong Kong
Phone: +852 829 48 24
Fax: +852 824 13 02
Telex: 5 01 51 INDHK HX

Hungary—MSZH ♦

Magyar Szabványügyi Hivatal
Pf. 24
1450 Budapest 9
Hungary
Phone: +36 1 118 30 11
Fax: +36 1 118 51 25
Telex: 22 57 23 NORM H

Iceland—STRI

Icelandic Council for Standardization
Technological Institute of Iceland
Keldnaholt
IS-112 Reykjavík
Iceland
Phone: +354 1 68 70 00
Fax: +354 1 68 74 09
Telex: 30 20 ISTECH IS

India—BIS ♦

Bureau of Indian Standards
Manak Bhavan
9 Bahadur Shah Zafar Marg
New Delhi 110002
India
Phone: +91 11 331 79 91
Fax: +91 11 331 40 62
Telex: 316 58 70 BIS IN

Indonesia—DSN ♦

Dewan Standardisasi Nasional—DSN (Standardization Council of Indonesia)
Sasana Wisya Sarwono Lt. 5
Jalan Jend. Gatot Subroto 10
Jakarta 12710
Indonesia
Phone: +62 21 520 66 74
Fax: +62 21 520 72 26
Telex: 6 28 75 PDII IA

Iran, Islamic Republic of—ISIRI ♦

Institute of Standards and Industrial Research of Iran
Ministry of Industry
P.O. Box 15875-4618
Tehran
Iran
Phone: +98 21 89 93 08
Fax: +98 21 89 53 05
Telex: 21 27 96 INMI IR

Iraq—COSQC

Central Organization for Standardization and Quality Control
Ministry of Planning
P.O. Box 13032
Aljadiria, Baghdad
Iraq
Phone: +964 1 776 51 80
Fax: +964 1 776 57 81
Telex: 21 35 05 COSQC

Ireland—NSAI ♦

National Standards Authority of Ireland
Glasnevin
Dublin 9
Ireland
Phone: +353 1 37 01 01
Fax: +353 1 36 98 21
Telex: 3 25 01 IIRS EI

Israel—SII ♦

Standards Institution of Israel
42 Chaim Levanon Street
Tel Aviv 69977
Israel
Phone: +972 3 54 54 154
Fax: +972 3 641 96 83
Telex: 3 55 08 SIIT IL

Italy—UNI ♦

Ente Nazionale Italiano di Unificazione
Via Battistotti Sassi 11
I-20133 Milano
Italy
Phone: +39 2 70 02 41
Fax: +39 2 70 10 61 06
Telex: 31 24 81 UNI I

Jamaica—JBS ♦

Jamaica Bureau of Standards
6 Winchester Road, P.O. Box 113
Kingston 10
Jamaica
Phone: +500 926 3140-6
(809) 926 3140-6 (from U.S.A. or Canada)
Fax: +500 921 5329
(809) 921 5329 (from U.S.A. or Canada)
Telex: 22 91 STANBUR

Japan—JISC ♦

Japanese Industrial Standards Committee
c/o Standards Department Agency of Industrial Science and Technology
Ministry of International Trade and Industry
1-3-1, Kasumigaseki, Chiyoda-ku
Tokyo 100
Japan
Phone: +81 3 35 01 92 95/6
Fax: +81 3 35 80 14 18
Telex: 02 42 42 45 JSATYO J

Jordan ✉

Directorate of Standards and Measures
Ministry of Industry and Trade
P.O. Box 2019
Amman
Jordan
Phone: +962 6 66 31 91
Fax: +962 6 60 37 21
Telex: 2 11 63 MINTR JO

Kenya—KEBS ♦

Kenya Bureau of Standards
Off Mombasa Road
Behind Belle Vue Cinema
P.O. Box 54974
Nairobi
Kenya
Phone: +254 2 50 22 10/19
Telex: 2 52 52 VIWANGO

Korea (North)—CSK

Committee for Standardization of the Democratic People's Republic of Korea
Zung Gu Yok Seungli-Street
Pyongyang
Democratic People's Republic of Korea
Phone: 57 15 76
Telex: 59 72 TECH KP

Korea (South)—KBS ♦

Bureau of Standards Industrial Advancement Administration
2, Chungang-Dong Kwachon-city
Kyonggi-do 427-010
Republic of Korea
Phone: +82 2 503 79 28
Fax: +82 2 503 79 41
Telex: 2 84 56 FINCEN K

Kuwait ✉

Standards and Metrology Department
Ministry of Commerce and Industry
Post Box No. 2944 Safat
13030 Kuwait
Fax: +965 242 44 11
Telex: 2 26 82 COMMIND KT

Libya (Libyan Arab Jamahiriya)—LNCSM

Libyan National Centre for Standardization and Metrology
Industrial Research Centre Building
P.O. Box 5178
Tripoli
Libya
Phone: +218 21 469 37
Fax: +218 21 469 37
Telex: 205 49 NCSM

Madagascar ✉

Direction qualité et métrologie légale
B.P. 1316
101 Antananarivo
Madagascar
Phone: +261 2 238 60
Telex: 22 378 MIN CO MG

Malawi ✉

Malawi Bureau of Standards
P.O. Box 946
Blantyre
Malawi
Phone: +265 67 04 88
Fax: +265 67 07 56
Telex: 4 43 25

Malaysia—SIRIM

Standards and Industrial Research Institute of Malaysia
Persiaran Dato' Menteri, Section 2
P.O. Box 7035
40911 Shah Alam
Selangor Darul Ehsan
Malaysia
Phone: 60 3 559 26 01
Fax: 60 3 550 80 95
Telex: 3 86 72 MA

Mauritius ✉

Mauritius Standards Bureau
Ministry of Industry and Industrial Technology
Reduit
Mauritius
Phone: +230 454 19 33
Fax: +230 464 76 75
Telex: 42 49 EXTERN IW

Mali ✉

Direction nationale des Industries de Mali
Ministère de l'Economie et des Finances
B.P. 278
Bamako
Mali
Phone: +223 22 57 56
Fax: +223 22 88 53
Telex: 25 59 MJ

Malta ✉

Malta Board of Standards
Department of Industry
St. George's, Cannon Road
Santa Venera
Malta
Phone: +356 44 62 50
Fax: +356 44 62 57

Mexico—DGN ♦

Dirección General de Normas
Calle Puente de Tecamachalco Nº 6
Lomas de Tecamachalco
Sección Fuentes
Naucalpan de Juárez
53 950 Mexico
Phone: +52 5 520 84 94
Fax: +52 5 540 51 53
Telex: 177 58 40 IMCEME

Mongolia—MNIS

Mongolian National Institute for Standardization
Ulaanbaatar 37
Mongolia
Phone: 3 29 30
Telex: 72233 (MNIS) MN

Morocco—SNIMA

Service de normalisation industrielle marocaine
1, Place Sefrou (Tour Hassan)
Rabat
Morocco
Phone: +212 72 45 30
Telex: 3 18 72

Nepal ✉

Nepal Bureau of Standards and Metrology
B.P. 985
Sundhara, Kathmandu
Nepal
Phone: +977 1 27 26 89

Netherlands—NNI ♦

Nederlands Normalisatie-instituut
Kalfjeslaan 2
P.O. Box 5059
N-2600 GB Delft
Netherlands
Phone: +31 15 69 01 90
Fax: +31 15 69 01 90
Telex: 3 81 44 NNI NL

New Zealand—SANZ ♦

Standards Association of New Zealand
Private Bag
Wellington
New Zealand
Phone: +64 4 384 21 08
Fax: +64 4 384 39 38
Telex: 38 50 SANZ NZ

North Korea

[see *Korea (North)*]

Norway—NSF ♦

Norges Standardiseringsforbund
Postboks 7020 Homansbyen
N-0306 Oslo 3
Norway
Phone: +47 2 46 60 94
Fax: +47 2 46 44 57
Telex: 1 90 50 NSF N

Oman, Sultanate of ✉

Directorate General for Specifications and Measurements
Ministry of Commerce and Industry
P.O. Box 550
Muscat
Oman
Phone: +968 70 47 83
Fax: +968 79 59 92
Telex: 36 65 WIZARA ON

Pakistan—PSI ♦

Pakistan Standards Institution
39 Garden Road, Saddar
Karachi 74400
Pakistan
Phone: +92 21 772 95 27
Fax: +92 21 772 95 27

Papua New Guinea ✉

National Standards Council
P.O. Box 3042
Boroko
Papua New Guinea
Phone: +675 27 21 02
Fax: +675 25 24 03

Philippines—BPS ♦

Bureau of Product Standards
Department of Trade and Industry
361 Senator Gil J. Puyat Avenue
Makati
Metro Manila 3117
Philippines
Phone: +63 2 818 57 01
Fax: +63 2 817 98 70
Telex: 1 48 30 MTI PS

Poland—PKNMiJ ♦

Polish Committee for Standardization, Measures and Quality Control
Ul. Elektoralna 2
00-139 Warszawa
Poland
Phone: +48 22 20 54 34
Fax: +48 22 20 83 78
Telex: 81 36 42 PKN PL

Portugal—IPQ ♦

Instituto Português da Qualidade
Rua José Estêvão, 83-A
P-1199 Lisboa Codex
Portugal
Phone: +351 1 52 39 78
Fax: +351 1 53 00 33
Telex: 1 30 42 QUALIT P

Romania—IRS

Institut roumain de normalisation
13, rue Jean-Louis Calderon
Code 70201
Bucuresti 2
Romania
Phone: +400 11 14 40
Fax: +400 12 08 23
Telex: 1 13 12 IRS RO

Russian Federation—GOST

State Committee for Standardization, Metrology and Certification
Leninski Prospekt 9
Moskva 117049
Russian Federation
Phone: +7 095 236 40 44
Fax: +7 095 236 82 09
Telex: 41 13 78 GOST SU

Saudi Arabia—SASO ♦

Saudi Arabian Standards Organization
P.O. Box 3437
Riyadh 11471
Saudi Arabia
Phone: +966 1 479 30 46
Fax: +966 1 479 30 63
Telex: 40 16 10 SASO SJ

Seychelles, Republic of ✉

Department of Industry
P.O. Box 648
Bel Eau, Mahe
Republic of Seychelles
Phone: +248 2 50 60
Fax: +248 2 50 86
Telex: 24 22 IND SZ

Singapore— SISIR ♦

Singapore Institute of Standards and Industrial Research
1 Science Park Drive
Singapore 0511
Phone: +65 778 77 77
Fax: +65 778 00 86
Telex: RS 2 84 99 SISIR

Slovenia—SMIS

Standards and Metrology Institute of Slovenia
Ministry of Science and Technology
Slovenska 50
61000 Ljubljana
Slovenia
Phone: +38 61 111 107
Fax: +38 61 124 288

South Africa, Republic of—SABS ♦

South African Bureau of Standards
Private Bag X191
Pretoria 0001
Republic of South Africa
Phone: +27 12 428 79 11
Fax: +27 12 344 15 68
Telex: 32 13 08 SA

South Korea

[see *Korea (South)*]

Spain—AENOR ♦

Asociación Española de Normalización y Certificación
Calle Fernández de la Hoz, 52
E-28010 Madrid
Spain
Phone: +34 1 410 48 51
Fax: +34 1 410 49 76
Telex: 4 65 45 UNOR E

Sri Lanka—SLSI ♦

Sri Lanka Standards Institution
53 Dharmapala Mawatha
P.O. Box 17
Colombo 3
Sri Lanka
Phone: +94 22 60 51
Fax: +94 1 44 60 18

Sweden—SIS ♦

SIS - Standardiseringskomissionen i Sverige
Box 3295
S-103 66 Stockholm
Sweden
Phone: +46 8 613 52 00
Fax: +46 8 11 70 35
Telex: 1 74 53 SIS S

Switzerland—SNV ♦

Swiss Association for Standardization
Kirchenweg 4
CH-8032 Zürich
Switzerland
Phone: +41 1 384 47 47
Fax: +41 1 384 47 74
Telex: 75 59 31 SNV CH

Syria—SASMO ♦

Syrian Arab Organization for Standardization and Metrology
P.O. Box 11836
Damascus
Syria
Phone: +963 11 45 05 38
Telex: 41 19 99 SASMO

Tanzania—TBS ♦

Tanzania Bureau of Standards
P.O. Box 9524
Dar es Salaam
Tanzania
Phone: +255 51 4 80 51
Fax: +255 51 4 80 51
Telex: 4 16 67 TBS TZ

Thailand—TISI ♦

Thai Industrial Standards Institute
Ministry of Industry
Rama IV Street
Bangkok 10400
Thailand
Phone: +66 2 245 78 02
Fax: +66 2 247 87 41
Telex: 8 43 75 MINIDUS TH (attention TISI)

Trinidad and Tobago—TTBS ♦

Trinidad and Tobago Bureau of Standards
P.O. Box 467
Port of Spain
Trinidad
Phone: +500 662 8827
809 662 8827 (from U.S.A. and Canada)
Fax: +500 663 4335
809 663 4335 (from U.S.A. and Canada)

Tunisia—INNORPI

Institut national de normalisation et de la propriété industrielle
B.P. 23
1012 Tunis-Belvédère
Tunisia
Phone: +216 1 78 59 22
Fax: +216 1 78 15 63
Telex: 1 36 02 INORPI TN

Turkey—TSE ♦

Türk Standardlari Enstitüsü
Necatibey Cadessi 112
Bakanliklar
06100 Ankara
Turkey
Phone: +90 4 117 83 30
Fax: +90 4 125 43 99
Telex: 4 20 47 TSE TR

Uganda ✉

Uganda National Bureau of Standards
P.O. Box 6329
Kampala
Uganda
Phone: +256 41 25 86 69

United Arab Emirates ✉

Directorate of Standardization and Metrology
P.O. Box 433
Abu Dhabi
United Arab Emirates
Phone: +971 2 72 60 00
Fax: +971 2 77 33 01
Telex: 2 29 37 FEDFIN EM

United Kingdom—BSI ♦

British Standards Institution
2 Park Street
London W1A 2BS
England
Phone: +44 71 629 9000
Fax: +44 71 629 0506
Telex: 26 69 33 BSILON G

United States of America—ANSI ♦

American National Standards Institute
Sales Department
11 West 42nd Street
New York, NY 10036
U.S.A.
Phone: (212) 642-4900
Fax: (212) 398-0023
(212) 302-1286 (Sales)
Telex: 42 42 96 ANSI UI
Hours: 8:45 to 16:45 Eastern Time

Publications

Catalog of American National Standards: $20, plus $4 (in the U.S.) for shipping and handling.
ISO Catalogue: $34, plus $5 (in the U.S.) for shipping and handling.
Payment terms: Check or money order (US$ only) with order. Billing against a purchase order is available only to ANSI members.

Uruguay ✉

Instituto Uruguayo de Normas Técnicas
San José 1031 P.7, Galería Elysée
Montevideo
Uruguay
Phone: +598 2 91 20 48
Fax: +598 2 92 16 81
Telex: 2 31 68 ANCAP B UY

Venezuela—COVENIN ♦

Comisión Venezolana de Normas Industriales
Avenida Andrés Bello
Edificio Torre Fondo Común, Piso 12
Caracas 1050
Venezuela
Phone: +58 2 575 22 98
Fax: +58 2 574 13 12
Telex: 2 42 35 MINFO VC

Viet Nam, Socialist Republic of—TCVN

General Department for Standardization, Metrology and Quality
70, Tran Hung Dao Street, Box 81
Hanoi
Viet Nam
Phone: +84 4 25 63 75
Fax: +84 8 9 30 12
Telex: 41 22 87 UKKN VT

Yugoslavia—SZS ♦

Savezni zavod za standardizaciju
Slobodana Penezica Krcuna br. 35
Post. Pregr. 933
YU-11000 Beograd
Yugoslavia

Phone: +38 11 64 40 66
Fax: +38 11 235 10 36
Telex: 1 20 89 JUS YU

Zimbabwe—SAZ

Standards Association of Zimbabwe
17 Coventry Road
Harare
Zimbabwe

Phone: +263 4 70 60 52

PUBLISHERS AND RESELLERS OF STANDARDS DOCUMENTATION

Asia (South & East) and Australasia

Global Info Centre
Suite 1310, Tower 1
World Trade Square
123 Hoi Bun Road
Kowloon
Hong Kong

Phone: +852 755 6733
Fax: +852 756 4141

The Institution of Radio & Electronics Engineers of Australia
Commercial Unit 3
2 New McLean Street
Edgecliff
Sydney, NSW 2027
Australia

Phone: +61 2 327 4822
Fax: +61 2 362 3229

Canada

Electrical and Electronic Manufacturing Association of Canada (EEMAC)
Standards Sales Office
10 Carlson Court, Suite 500
Rexdale, Ontario M9W 6L2
Canada
Phone: (416) 674-7410
Fax: (416) 674 7412

EEMAC also accepts Canadian orders for hard copy EIA documents. They maintain an EIA document inventory in Canada.

Europe and Israel

American Technical Publishers Ltd.
27-29 Knowl Piece, Wilbury Way
Hitchin, Herts SG4 OSX
England
Phone: +44 462 37933
Fax: +44 462 33678
Telex: 825 648 ATPG

Europe, Middle East, and Africa

Global Info Centre
31-35, rue de Neuilly
92110 Clichy
France
Phone: +33 1 40 87 17 02
Fax: +33 1 40 87 07 52

Japan

Japanese Standards Association
1-24, Akasaka, 4-Chome
Minato-Ku
Tokyo 107
Japan
Phone: +81 3 583 8001
Fax: +81 3 586 2014

Maruzen Company Ltd.
3-10 Nihonbashi, 2-Chome
Chuo-Ku
Tokyo 103
Japan

This is the official Japanese agent for IEEE documents.

Latin America

Global Info Centre
1108 Normandy Drive
Miami Beach, FL 33141
U.S.A.
Phone: +1 305 868-2872
Fax: +1 305 864-4641

United States

Electronic Industries Association (EIA)
Standards Sales Office
2001 Pennsylvania Avenue, NW
Washington, DC 20006
Phone: (202) 457-4966
Fax: (202) 457-4985
TWX: 710-822-0148
Hours: 9:00 to 16:30 Eastern Time

Publications: EIA and JEDEC Standards.
Catalogue: $10.
Payment terms: Check or money order (US$ only) with order. Credit card (AE, MC and VISA) telephone orders (subject to surcharge—greater of $5 or 15% of order value). Sales tax applies to orders from Washington, DC. Normal turnaround is 24 to 48 hours. Expedited shipping is available at extra cost.

Global Engineering Documents
Phone: (800) 854-7179 (connects to the nearest U.S. regional office)

Eastern Region Office
1990 M Street NW, Suite 401
Washington, DC 20036
Phone: (202) 429-2860
Fax: (202) 331-0960
Hours: 8:00 to 16:30 Eastern Time

Central Region Office
7730 Carondolet Avenue, Suite 407
Clayton, MO 63105
Phone: (314) 726 0444
Fax: (314) 726-0618
Hours: 8:00 to 16:30 Central Time

Pacific Region Office
2805 McGaw Avenue, P.O. Box 19539
Irvine, CA 92714
Phone: (714) 261-1455
Fax: (714) 261-7892
Hours: 8:00 to 16:30 Pacific Time

Offers same-day service on U.S. orders for hard copy documents of most national and international standards.

National Standards Association
1200 Quince Orchard Boulevard
Gaithersburg, MD 20878
Phone: (301) 590-2300
Fax: (301) 990-8378

Offers individual standards or groups of standards, on microfiche, to U.S. and international customers.

Omnicom
501 Church Street NE, Suite 304
Vienna, VA 22180
Phone: (703) 281-1135
Fax: (703) 281-1505

Distributes ANSI and CCITT documents and publishes, annually, *The Omnicom Index of Standards.*

United Nations Bookshop
General Assembly Building
Room GA 32B
New York, NY 10017
Phone: (800) 553-3210 (U.S. outside NY state)
(212) 963-7680 (New York or international)
Hours: 9:30 to 17:45 Eastern Time, Monday to Saturday; 10:00 to 17:45 Eastern Time, Sunday.
Publications: CCITT documents.
Catalogue ("List of CCITT Blue Availables"): Free.
Payment terms: Check or money order (US$ only) or, if over $15, VISA or MasterCard.
Shipping charge (UPS) of $1 per book (free for more than 5 books).

Worldwide

Information Handling Services
P.O. Box 1154
15 Inverness Way East
Englewood, CO 80150
Phone: (800) 525-7052 (U.S., outside Colorado)
(303) 790-0600 (Colorado and international)

Specializes in microfiche, 16-mm microfilm, and CD-ROM annual subscription services, with updates every two months.

Section

3

IBM Communication Products

The following definitions include only products that are not industry-specific (e.g., banking terminals).

IBM 2780 Obsolete IBM remote-job-entry terminal. The IBM 2780 is used, via a BSC connection, as a remote card reader and printer for an IBM host computer. Personal computers are sometimes used for emulated 2780 operation, using files on disk rather than punched cards.

IBM 3101 ASCII Display Station IBM's first ASCII display terminal, introduced in 1980. No longer manufactured.

IBM 3102 Printer Companion printer to the 3101 ASCII Display Station. No longer manufactured.

IBM 3151 ASCII Display Station Depending on model, can provide character-mode, echo-mode or block-mode operation with screen format (row x column) of 24x80, 25x80, 24x132, 25x132, or 28x132. Operates at start-stop data rates up to 38.4 kbit/s. Using pluggable cartridges, provides emulation of the most popular ADDS, Data General, DEC, TeleVideo, Wyse, and Zentec (ex-Lear-Siegler ADM series) ASCII display terminals. Supports character sets defined by IBM Code Page 437 (U.S.) and 850 (international). Other model-dependent characteristics include keyboard (84, 101, or 102 keys) and phosphor color (green or amber-gold). All models have a 14-in monitor.

IBM 3161, 3162, and 3163 ASCII Display Stations 12-in (3161 and 3163) and 14-in (3162) lower-cost variants of 3151 Display Station. No longer manufactured.

IBM 3164 Color ASCII Display Station 14-in color display station, available in four models, of which two provide ALA (American Library Association) character support. One of the two ALA models supports the combinations of alphabetic characters and diacritical marks (accents, etc.) used in 25 languages. Can operate in character, echo, or block mode. Supports speeds up to 19.2 kbit/s via a CCITT V.24/V.28 (EIA-232-D) or V.24/V.11 (EIA-422-A) interface. Supports eight foreground and eight background colors, plus a number of display and data-entry attributes. Character set is 128-character ASCII (ISO 646), plus 24 line-drawing graphics, 10 superscript, and 10 subscript characters. The display format is 24x80.

IBM 3172 Interconnect Controller Micro Channel-based intelligent controller, used to attach Token-Ring, MAP 3.0, or Ethernet local area networks (LANs) to the channel of a System/370 or System/390 host processor. May also be used for remote channel-to-channel connection over a T1 link (1.544 Mbit/s). The 3172 Model 2, which uses an Intel 80486 processor, also supports the 100-Mbit/s Fiber Distributed Data Interface (FDDI). Host channel connection may be by means of System/370-type parallel channels or ESCON fiber optic channels. In conjunction with AT&T Paradyne, IBM is expected to introduce a model, sometime in 1992, supporting remote channel-to-channel connection over a T3 link (44 Mbit/s).

IBM 3174 Establishment Controller IBM 3270 Information Display System controller, capable of supporting up to 32 (64, with expansion feature, on some models) Category A coaxially connected devices (in CUT mode or DFT mode, single and multiplexed), Token-Ring devices, and up to 24 ASCII terminals. Depending on the model, the 3174 may communicate with the host system via a parallel or ESCON channel attachment (at 1.25 or 2.5 Mbyte/s), a remote communications link (up to 64 kbit/s), or via Token Ring (4 or 16 Mbit/s). Remote attachment options include CCITT V.24/V.28 (EIA-232-D), CCITT X.21, and CCITT V.35 interfaces and SNA/SDLC, CCITT X.25, and BSC operation. May also communicate, in start-stop mode, with an ASCII host.

IBM 3174 Subsystem Control Unit Successor to the IBM 3274 Cluster Controller and predecessor to the 3174 Establishment Controller. All models (1L, 1R, 2R, 3R, 51R, 52R, 53R, 81R, and 82R) were withdrawn in July 1989.

IBM 3270 Information Display System Terminal system consisting of cluster or establishment controllers, keyboard/display terminals, printers, etc.

Many individual IBM 3270 Information Display System devices are defined in *Sec. 4*.

IBM 3270 Workstation Program IBM software product which provides much of the capability of the IBM 3270 PC on a variety of IBM PCs, including especially IBM PS/2s.

IBM 3274 Cluster Controller 3270 series cluster controller, capable of supporting Category A coaxially connected devices (in CUT mode or DFT mode, single and multiplexed). Depending on the model, the 3274 may communicate with the host system via a channel attachment (at 640 kbyte/s), or a remote SDLC or BSC communications link. Early models of the IBM 3274 can also support Category B coaxially connected devices. It has been replaced by the 3174 Establishment Controller.

IBM 3278 Monochrome keyboard/display control-unit terminal. Although no longer manufactured, the 3278 is used as a reference point in describing the functionality of many 3270 terminal emulation programs. In particular, the screen formats supported by emulators correspond to those of four 3278 models, specifically:

3278 Model	Screen Format (Row x Col)
2	24 x 80
3	32 x 80
4	43 x 80
5	27 x 132

For all formats, there is an additional, status line (Operator Information Area—OIA).

For general specifications of this and other display terminals of the IBM 3270 Information Display System, see *IBM 3270 Display Terminals* in *Sec. 4*.

IBM 3287 Coaxially attached dot-matrix printer. The 3287 was available as Models 1 and 2 (monochrome) and as Models 1C and 2C (color). Although IBM offers a wide choice of coaxially attached printers, the 3287 is used as the reference point for host-addressable non-IPDS printer support in 3270 emulation products, even where the printer actually used is other than dot-matrix. See *SCS* and *IPDS* in *Sec. 4*.

For general specifications of this and other IBM 3270 Information Display System printers, see *IBM 3270 Printers* in *Sec. 4*.

IBM 3299 Terminal Multiplexer Multiplexer for up to 8 (Model 2 or 3) or up to 32 (Model 032) Category A devices. The 3299 may be connected via coaxial or twisted-pair cable to the terminals and to the terminal adapter of an IBM 3174, to a multiplexed port on an IBM 3274 (Models 2 and 3 only), or to the multiplexed port of another 3299. The 3299 Model 032 also supports a fiber optic connection to the 3174.

IBM 3705 Communication Controller IBM host-attached communication controller. With its associated software (ACF/NCP), the 3705 controls the operation of a number of attached communications lines, which may be operating in a variety of modes and under a variety of protocols. The 3705 is no longer manufactured but is often used as a nonspecific reference for compatible communication controllers.

IBM 3708 Network Conversion Unit Device providing concurrent line concentration, protocol conversion, protocol enveloping, and ASCII pass-through for start-stop ASCII devices. ASCII terminals can communicate with a host system in either native mode or as full-screen 3270 displays or printers. ASCII terminal, ASCII host, and SNA/SDLC host connections operate at up to 19.2 kbit/s. For ASCII to 3270 protocol conversion, its appearance to an SNA/SDLC host is the same as that of a 3274 Model 51C or 61C Cluster Controller.

IBM 3710 Network Controller Device providing the following concurrent functions:

- ASCII (start-stop) to SNA 3270 protocol conversion
- BSC 3270 to SNA 3270 protocol conversion
- Asynchronous and BSC (RJE) protocol enveloping
- ASCII pass-through
- Concentration of ASCII, BSC, and SNA/SDLC traffic

Host communication is over one or more user-defined SNA/SDLC or X.25 links.

IBM 3720 Communication Controller Low-capacity successor to the IBM 3705. Supports up to 28 communication lines (60 with the optional IBM 3721 Expansion Unit). This controller is no longer manufactured.

IBM 3725 Communication Controller Medium-capacity successor to the IBM 3705. This controller is no longer manufactured.

IBM 3745 Communication Controller Low-, medium-, and high-capacity successor to the IBM 3705. Depending on model and expansion options, supports from 16 to 896 communication lines and from 1 to 16 host system attachments. Some models are designed for remote operation, as line concentrators.

IBM 3780 IBM remote-job-entry terminal. The IBM 3780 is used, via a BSC or SDLC connection (3780 model-dependent), as a remote card reader and printer for an IBM host computer. Personal computers are often used for emulated 3780 operation, using ASCII files on disk rather than punched cards. IBM 3780 operation is supported by IBM's Job Entry System (JES).

IBM 37x5 (or 37xx) Generic term for IBM 3705, 3720, 3725, or 3745. Without the IBM prefix, may also refer to fully compatible non-IBM equivalents.

IBM 3x74 Generic term for IBM 3174 or 3274.

IBM 4361 Workstation Adapter Built-in IBM 3x74 cluster controller in an IBM 4361 host system (no longer manufactured). The Workstation Adapter has four terminal adapter ports to which IBM 3299s or equivalents may be connected.

IBM 5208 ASCII-5250 Link Protocol Converter Device allowing ASCII displays, PCs emulating ASCII displays, and ASCII printers to attach, as 5250 terminals, to the workstation controller (twinaxial) of a System/36, System/38, or AS/400 or to a 5294 or 5394 Remote Controller. Up to seven ASCII devices, which may be local or remote, are supported. One parallel printer is also supported.

IBM 5209 Link Protocol Converter Device allowing up to seven 3270 displays and/or printers to be attached to a System/36, System/38, or AS/400 or to a 5294 or 5394 Remote Controller, and concurrently to an IBM 3174 or 3274 controller. 3270 displays can be "hot-keyed" between 3270 and 5250 sessions. 3270 printers can be dynamically assigned as either 3270 or 5250 printers. To the 3174 or 3274 controller, the 5209 appears as a 3299 Terminal Multiplexer.

IBM 5250 Information Display System Terminal system, consisting of keyboard/display terminals and printers, supported by IBM System/34, System/36, and System/38 host minicomputers. Interterminal connections and direct host connections are made by means of twinaxial cable. Newer 5250-class devices (e.g., IBM 5294 controller, IBM 3197 Model C, and IBM 3197 Model D terminals) also support twisted-pair cabling. Note that the newer 5250-class display terminals have 31xx or 34xx designations.

IBM 5251 IBM monochrome keyboard/display terminal for twinaxial connection to a System/34, System/36, or System/38 minicomputer, or to another IBM 5251 with a built-in controller function. The 5251 Model 12, no longer manufactured, includes the controller function; the 5251 Model 11 does not. Although also no longer manufactured, the IBM 5251 Model 11 is used a reference point in describing single-station 5250 emulation products which use a twinaxial interface.

For general specifications of this and other display terminals of the *IBM 5250 Information Display System,* see *Sec. 4.*

IBM 5294 Remote SDLC control unit, supporting up to eight IBM 5250 devices via twinaxial or twisted-pair cable. Communicates with an IBM System/36 or System/38. It is no longer manufactured, having been superseded by the IBM 5394 and 5494.

IBM 5299 Terminal Multiconnector and Twinax to Telephone Twisted Pair (TTP) Adapter. Connects up to seven 5250-type terminals to a System/36, System/38, or AS/400 or a 5294 or 5394 remote SDLC control unit. Terminal-to-5299 connections are by means of Telephone Twisted Pair cable. 5299-to-host connections are by means of TTP or twinaxial cable.

IBM 5394 Remote SDLC, X.21, or X.25 control unit, supporting up to 8 IBM 5250 devices in 5294 emulation mode, or 16 devices in 5394 native mode, via twinaxial or twisted-pair cable. Communicates with an IBM System/36, System/38, or AS/400 as a Physical Unit (Node) Type 1. Only the AS/400 supports 5394 native mode.

IBM 5494 Remote SDLC, X.21, or X.25 control unit supporting up to 28 IBM 5250 devices via twinaxial or, using IBM 5299s, twisted-pair cable. Model 2 also supports a Token-Ring, either for the host link or as a gateway for up to 40 workstations. Communicates with an IBM AS/400 as a Physical Unit (Node) Type 2.1.

IBM 8775 Display Terminal 3270 data-stream-compatible terminal, with loop attachment to IBM 4331 or 8100 or SDLC attachment to 4300 series, System/370, System/390, and 8100 systems.

Models 1 and 2 support loop attachment at speeds up to 38.4 kbit/s. Models 11 and 12 support SDLC at speeds up to 9600 bit/s. All models support 12x80, 24x80, and 32x80 screen formats. Models 2 and 12 also support a screen format of 43x80.

The 8775 is interesting in that it supports an Enhanced Function, which includes the Field Validation extension of Extended Field Attributes. This provides for Mandatory Enter, Mandatory Fill, and Trigger fields. See *attribute* in *Sec. 4.*

IBM TokenWay 3174 Establishment Controller Low-cost version (Model 90R) of the IBM 3174 Establishment Controller, supporting one Category A terminal adapter, one 4- or 16-Mbit/s Token-Ring network, and an SNA/SDLC host communication link at up to 64 kbit/s. In conjunction with an IBM 3299 Terminal Multiplexer, the terminal adapter can support up to eight 3270-type display terminals or printers.

TIC Token-Ring Interface Coupler. Device on IBM 3725 communication controller, allowing the use of Token-Ring topology for communication with local terminal emulation devices. IBM 3720 and 3745 communication controllers also support Token-Ring attachment via one (3720) or up to four (3745) Token-Ring Adapters.

Models 1 and 2 support loop attachment at speeds up to [illegible]. The [illegible] and [illegible] support SDLC [illegible] at [illegible] bps. All models support [illegible] 2[illegible] [illegible] 30 [illegible] [illegible] formats. Models 2 [illegible] 12 [illegible] screen [illegible] the [illegible] [illegible] format [illegible] [illegible] which includes the Field Validation [illegible] and [illegible] Field Attributes. This provides [illegible] [illegible] [illegible] Fill and Trigger Fields. [illegible]

IBM Token-Ring [illegible] Establishment Controller [illegible] Model 3[illegible] of the [illegible] [illegible] [illegible] [illegible] Category [illegible] [illegible] [illegible] [illegible] and [illegible] [illegible] [illegible] to 8[illegible] bytes. In conjunction with [illegible] [illegible] [illegible] [illegible] [illegible] [illegible] [illegible] display [illegible]

TIC [illegible] Interface [illegible] [illegible] IBM [illegible] communication controller [illegible] the use of [illegible] for communication with [illegible] station [illegible]. IBM 3720 and 3725 communication [illegible] [illegible] [illegible] [illegible] Attachment [illegible] (37[illegible]) [illegible]

Section

4

IBM Information Display Systems

The IBM 3270 Information Display System

The IBM 3270 Information Display System (3270 IDS) is a key component of Systems Network Architecture (SNA), as implemented in a hierarchical (host-based) network.

Interactive terminal-host communication, under SNA, is based on Physical Unit Type 2 (Peripheral Node—usually a 3274, 3174, or equivalent), Logical Unit Type 2 (3270 display station or equivalent), Logical Unit Type 3 (3270 printer), and Logical Unit Type 1 (SCS printer). Micro-to-mainframe communication depends on the emulation of some or all of these Network-Addressable Units (NAUs) on personal computers and similar devices.

In spite of (or reinforced by) the enormous success of 3270-based emulation products, both in standalone and gateway-based LAN-resident forms, IBM continues to enhance and expand the 3270 IDS itself, offering it to a market which is large enough to accommodate at least three successful imitators (IDEA Courier, McDATA, and Memorex Telex).

The 3174 Establishment Controller's newly expanded capability, as a Physical Unit Type 2.1 (peer-to-peer networking node), restricts future growth in the use of 3270 displays and printers. However, the large investment in host software which supports the 3270 data stream indicates that understanding and support of the existing architectures will continue to be important for several years.

This section attempts to cover most of the terminology which might be encountered in dealing with the 3270 Information Display

System itself, with compatible equivalents, and with emulation products. It does not deal with terminology applicable to the non-3270 environment (e.g., a local area network) in which 3270 operation might be emulated.

IBM 3270 Information Display System Terminology

active logical terminal For multiple logical terminal (MLT) support, the logical terminal currently displayed. Also called *foreground logical terminal*. See also *background logical terminal*.

AEA Asynchronous Emulation Adapter. Feature of the 3174 Establishment Controller, which provides full-duplex, start-stop mode communication at up to 19.2 kbit/s to up to eight ASCII display terminals, printers, and/or host systems. Communication is character (conversational) mode only (i.e., there is no half-duplex block-mode support). Depending on the model, the 3174 can accommodate either one or up to three AEAs. The following logical connections are supported:

- ASCII display terminal or ASCII printer to ASCII host
- ASCII display terminal, emulating 3270 display terminal (with 24x80 or 32x80 screen format), to IBM host
- ASCII printer, emulating IBM 3287, to IBM host
- 3270 display terminal, concurrently emulating up to five ASCII display terminals, to up to five ASCII hosts (mixed IBM and ASCII host sessions may be supported on the same 3270 display terminal, to a maximum of five)
- 3270 printer (3287 Model 1 or 2, 3262 Model 3 or 13, 4224 Model 201 or 202, or 5204 in 3287 emulation mode), emulating an ASCII printer, to ASCII host

APL "A Programming Language." A high-level language, which uses an extended character set (the APL character set) and which is used extensively by the academic community and in the actuarial field (among others). Support of APL on 3270 terminals involves the use of extended data stream (EDS) and the extended attribute buffer (EAB).

APL2 An extended version of APL. Also, an extended version of the APL character set, having 10 more symbols than the basic set.

attribute Any of several characteristics of a field (one or more successive characters) or of a single character. Field Attributes are display, nondisplay (for the entry of hidden data, such as passwords), normal, intensified, protected (for display only), unprotected (for data entry), light-pen selectable, skip (automatically bypassed when tabbing), and numeric-only. Additional attributes which can be applied to a field (Extended Field Attribute) or a single character (Character Attribute) are color (any of 7), blink, reverse video, underline, and symbol set (e.g., basic 3270, APL, or any of 6 Programmed Symbol sets). For a detailed description, see *attribute byte*.

attribute byte A byte, within the device buffer, containing the attribute information for the bytes following it (Field Attribute byte). Alternatively, a byte, in an Extended Attribute Buffer (EAB), containing extended attribute information for the corresponding device-buffer Field Attribute byte or attribute information for the corresponding device-buffer character.

Bit Definitions in 3270 Field Attribute Byte

0	1	2	3	4–5	6	7
1	1	U/P	A/N	D/SPD	Res	MDT

Bits	Description
0, 1	In the 3270 device buffer, these 2 bits are set to 11, uniquely identifying the byte as a Field Attribute byte. In the 3270 data stream, they are set to a value which will make the byte an EBCDIC/ASCII translatable (displayable/-printable) graphic.
2 U/P	0 = Unprotected 1 = Protected
3 A/N	0 = Alphanumeric 1 = Numeric (or, in conjunction with a value of 1 for bit 2, Skip)
4, 5 D/SPD	00 = Display and not selector-lightpen detectable 01 = Display and selector-lightpen detectable 10 = Intensified display and selector-lightpen detectable 11 = Nondisplay (and nonprint) and not detectable
6	Reserved bit. Must always be zero.
7 MDT	Modified Data Tag; identifies modified fields for transmission to the host during Read-modified command operations. 0 = Not modified 1 = Modified Lightpen-selectable fields [detectable fields with a question mark (?) in the first byte] can be modified and unmodified with the lightpen or the cursor select key.

The Field Attribute byte occupies a buffer position, and hence a screen position, immediately to the left of the field to which it applies. It displays as a space.

An Extended Field Attribute byte and one or more character attribute bytes occupy positions in the Extended Attribute

Buffer (see *EAB*) corresponding to the device buffer positions occupied by the Field Attribute byte and the displayable bytes of the associated field. The following illustration shows this relationship.

3270 Field Attribute, Display Character, Extended Field Attribute (EFA), and Character Attribute (CA) Relationships

EAB	EFA	CA	CA	CA	CA	CA	CA
Device Buffer	Field Attribute	Character	Character	Character	Character	Character	Character

For most displays, the Extended Field Attribute byte contains from one to three fields.

Bit Definitions in 3270 Extended Field Attribute and Character Attribute Bytes

0	1	2	3	4	5	6	7
Extended Highlighting		Color			Symbol Set		
		Red	Green	Blue			

Bits	Description
0 - 1	**Extended Highlighting** 00 = No extended highlighting 01 = Underscore 10 = Blink 11 = Reverse video
2 - 4	**Color** 000 = None 001 = Blue 010 = Green 011 = Turquoise 100 = Red 101 = Pink 110 = Yellow 111 = White (Neutral)
5 - 7	**Symbol Set** 000 = Base (Symbol Set 0) 001 = APL/Text (Symbol Set 1) 010 to = Programmed Symbol (PS) Set (2 to 7) 111

Character Attribute bytes are the same as Field Attribute bytes, except that they apply to individual characters in the corresponding device buffer position. One exception is that, for each of the attribute types (Extended Highlighting, Color, Programmed Symbols), a value of zero causes the character attribute to default to the corresponding field attribute. This characteristic is called *Field Inherit.*

Monochrome terminals without programmed symbol capability only require the 2 bits for Extended Highlighting, plus 1 bit for Symbol Set selection, leaving 5 bits for other potential uses. The

IBM 8775 Display Terminal, for example, uses 3 bits for field validation—1 for Mandatory Fill, 1 for Mandatory Entry, and 1 to identify the field as a Trigger Field (to be sent to the host as soon as data has been entered and the cursor has been moved to another field).

attribute-select keyboard Keyboard available on some 3270 display terminals with the EAB (Extended Attribute Buffer), with key assignments for the entry of character attributes. Any character attribute (extended highlighting, programmed symbol set, if supported, or color) or valid combination can be selected, including field inherit.

background logical terminal For multiple logical terminal (MLT) support, any of the logical terminals not currently displayed. See also *active logical terminal*.

balun Balanced-to-unbalanced connector. Used to connect twisted-pair cable to a port on a Cluster/Establishment Controller, terminal or similar device. See also *coaxial cable*.

base color mode A mode in which color terminals (e.g., IBM 3279) represent four field attribute possibilities using arbitrary colors, specifically:

Attribute Combination	Color
Unprotected, normal	Green
Unprotected, intensified	Red
Protected, normal	Blue
Protected, intensified	White

Additional attributes (e.g., selector-lightpen detectable) do not cause any variation in these colors.

Category A Term referring to a coaxial connection, operating with an IBM-defined protocol at a data rate of 2.35 Mbit/s at a distance of up to 1500 m. All current IBM 3270 IDS coaxial terminals and controllers are classified as Category A. Recent Category A devices may also use twisted-pair cable.

Category B IBM's earlier coaxial connection protocol, operating over a shorter distance and at a much lower data rate than Category A. IBM 3271 and 3272 Cluster Controllers, along with 3277 display terminals and 3284 printers, were classified as Category B.

Character Attribute See *attribute, attribute byte,* and *EAB*.

Cluster Controller The IBM 3274, which is capable of supporting

Category A coaxially connected devices (in CUT mode or DFT mode, single and multiplexed). Depending on the model, the 3274 may communicate with the host system via a channel attachment (at 640 kbyte/s) or a remote SDLC or BSC communications link. Early models of the IBM 3274 can also support Category B coaxially connected devices. It has been replaced by the 3174 Establishment Controller.

coax Common abbreviation for coaxial cable.

coaxial Having a common axis.

coaxial cable Cable consisting of a center conductor (usually copper), surrounded by a braided tubular conductor of constant diameter. The two conductors are separated by a tubular polyethylene insulator and the cable is encased in a vinyl sheath. The electrical impedance of coaxial cables used for 3270 connections must be 93 Ω. In many cases, twisted-pair cable is supplanting the use of coaxial cable, with connection and impedance matching being handled at one or both ends by means of a balun (balanced to unbalanced) connector.

Control-Unit Terminal See *CUT*.

CUT Control-Unit Terminal. A terminal whose keystrokes are processed and whose presentation space (device buffer) is managed by the cluster controller to which it is attached (via Category A coaxial cable). A control unit terminal is capable of supporting only one host communication session at a time. (However, the IBM 3174 Establishment Controller will allow the operator of a control-unit terminal to switch among up to five host sessions, all of which are maintained by the controller.) Examples of control unit terminals are the IBM 3178, 3179, 3180, 3278, and 3279. Their mode of operation is referred to as *CUT mode*. IBM 3270 series printers (e.g., IBM 3287) also operate in CUT mode.

CUT mode See *CUT*.

device buffer The area of memory within a terminal in which the information that is to be displayed or printed is stored. Thus, a device buffer may be a screen buffer or a printer buffer.

DFT Distributed-Function Terminal. Category A coaxially connected device which does not require cluster-controller interaction to respond to keystrokes. A DFT terminal may contain one or more NAUs, typically LU Type 1, 2, or 3. Communication between a DFT and a cluster controller is in the form of EBCDIC message blocks. Coaxially connected 3270 gateways and multi-session and vector graphics coaxially connected terminals always operate in DFT mode. See also *CUT*.

DFT-E DFT Extended. DFT extension which allows the display terminal to be used for controller customization (normally only accessible by a control-unit terminal), provides access to an ASCII host session, and supports X.21/X.25 and controller diagnostics. The controller must be an IBM 3174 Establishment Controller or an appropriately featured equivalent.

EAB Extended Attribute Buffer. A second buffer containing additional attribute information for each position of the device buffer. Each position in the EAB corresponds to a position in the device buffer. EAB bytes corresponding to device-buffer attribute bytes define additional attributes for the associated field; such attributes are called *extended field attributes*. EAB bytes corresponding to device-buffer display characters define individual attributes for those characters, either superseding, modifying, or "inheriting" the field attributes. Color models of 3270 printers (e.g., IBM 3287 Model 1C) also have EABs. See *attribute* and *attribute byte*.

EDS Extended Data Stream. 3270 data stream containing information destined for the EAB (Extended Attribute Buffer).

eight-color support Similar to seven-color support, except that black is included as one of the colors. The term *eight-color support* is normally used where the screen background color can be other than black. Black, therefore, becomes one of the foreground color options (e.g., with graphics).

Entry Assist A customizable feature of IBM 3174, 3274, and equivalent cluster controllers which, in conjunction with a CUT-mode terminal, provides local control of margins, word wrap, end-of-line warnings, and tab stops. Several host applications (not recent ones) have been designed to take advantage of Entry Assist to facilitate the entry of continuous text.

Establishment Controller The IBM 3174, which is capable of supporting up to 32 (64, with expansion feature, on some models) Category A coaxially connected devices (in CUT mode or DFT mode, single and multiplexed), Token-Ring devices and a variety of ASCII terminals. Depending on the model, the 3174 may communicate with the host system via a parallel or ESCON channel attachment (at 1.25 or 2.5 Mbyte/s), a remote communications link (up to 64 kbit/s), or Token-Ring (4 or 16 Mbit/s).

Remote attachment options include CCITT V.24/V.28 (EIA-232-D), CCITT X.21, and CCITT V.35 interfaces, and SNA/SDLC, CCITT X.25, and BSC operation. (Establishment Controller is a comparatively recent term.)

Extended Field Attribute See *attribute, attribute byte,* and *EAB*.

Extended Graphics mode 3174 function which extends the number of characters that can be entered at the terminal keyboard. Can be used in 3270 mode or ASCII terminal emulation mode. The Alt-Shift key combination is used to indicate that the following keystroke is to be in Extended Graphics mode. See also *language shift*.

Extended Highlighting The attributes of reverse video, blink, and underline. See also *attribute byte*.

Extended Vital Product Data For IBM SNA devices, particularly terminals, machine type, model number, serial number, and an additional 320 bytes of data. Provides for NetView asset management. See *Vital Product Data*.

Field Attribute See *attribute* and *attribute byte*.

field inherit A property of Character Attributes in which a zero value for the color, extended highlighting, or symbol set attribute bit groups causes the associated character to "inherit" the corresponding attribute defined in the Field Attribute byte. See also *attribute* and *attribute byte*.

foreground logical terminal See *active logical terminal*.

host-addressable printer A printer, operating as an NAU, and capable, therefore, of accepting a data stream from a host computer.

input inhibited Display terminal status in which keystrokes are not processed. If input is inhibited while the terminal awaits a host response, keystrokes may be entered if the terminal supports the type ahead function. If input is inhibited because of an invalid operator action (e.g., an attempt to enter data into a protected field), it may be reenabled by means of the reset key.

inverse video See *reverse video*.

IPDS Intelligent Printer Data Stream. IBM's approach to host-initiated printer data streams. It uses structured fields (see *structured field*) to send data and commands to a printer, independently of the attachment protocol and, therefore, of the system to which the printer is attached.

Commands allow the host system to control and manage the downloading of fonts, symbol sets, and stored objects such as overlays and page segments. IPDS can be used for the presentation of text, raster images, vector graphics, bar codes, and previously stored overlays anywhere on a page.

As the use of sophisticated dot-matrix printers, laser printers and other advanced imaging devices increases, the use of IPDS is becoming widespread.

keyboard/display terminal Terminal capable of supporting one or more operator to host keyboard/display sessions. Some keyboard/display terminals (e.g., IBM InfoWindow 3472) can also support a host-addressable printer session using an attached printer.

language shift A function allowing a display terminal operator to select one of two supported character sets (e.g., Latin and Greek). Language switching may be implemented via a toggle function, with successive depressions of the function key (or key combination) switching back and forth or by means of designated shift functions (e.g., Latin Lower, Latin Upper, Greek Lower, Greek Upper).

local format storage 3174 Establishment Controller function which allows pre-formatted screens to be downloaded from the host system. Presentation of formats is subsequently invoked by the host application (CICS only). May be used only with control-unit terminals (see *CUT*).

logical address For a DFT-mode port of an IBM 3174 Establishment Controller or 3274 Cluster Controller, one of up to four addresses assignable in addition to the physical port address.

MLT Multiple Logical Terminals. Support for multiple logical units via a single terminal port on an IBM 3174 Establishment Controller, 3274 Cluster Controller, or equivalent. Depending on the level chosen at controller customization time, can support up to five display sessions on a control-unit terminal (3174 only) or up to five display and/or printer sessions, in any mix, on a distributed function terminal. See also *DFT*.

Model 2, 3, 4, 5 Although current IBM 3270 Information Display Systemterminal model numbers are not used to indicate screen format, these four 3278 model numbers are still often used, informally, to designate screen sizes. See also *IBM 3278* in *Sec. 3*.

multi-session The support of multiple communication sessions within one terminal or terminal-emulation device. See *MLT*.

national use Language-dependent usage, especially of character-set codes. There are, for example, 14 EBCDIC code points which represent different characters, depending on national language, and for which there are language-to-language differences in the corresponding 3270 buffer codes. See *EBCDIC* (in *Sec. 5*) and *buffer code*.

OIA Operator Information Area. Nondata area occupying the last row on a display terminal. Used to provide error and status information. Often referred to as the status line.

physical address For a cluster or establishment controller, the address associated with a physical (coaxial or twisted-pair) connection. When the controller is customized for DFT-mode operation, up to four logical addresses may be associated with each physical address. Controller ports to which an IBM 3299 may be connected have 8 or 32 (3299 Model 032) physical addresses, each of which can have 4 additional logical addresses.

presentation space The display or printer device buffer, corresponding to data displayed on the screen, including the Operator Information Area (OIA), or data to be printed. Because data entered at the keyboard is stored in the presentation space, display terminal presentation spaces can contain both received data and data to be transmitted.

printer authorization matrix (PAM) A 3174 Establishment Controller feature which allows printers or classes of printers to be associated with display terminals for the purpose of local-copy printing (screen printing).

programmed symbol A symbol which is created by the host graphics manager [for example, IBM's GDDM (Graphical Data Display Manager)] from an area, within a bit-mapped graphical image in memory, having the same dimensions (horizontal PELs by vertical PELs) as the displayable characters on the target terminal. Programmed symbols are transmitted, in structured fields, for storage in the terminal's "symbol-set memory" and subsequent display. For each character position on the screen, the appropriate stored programmed symbol is selected by means of a symbol set number in the EAB and a value, in the device buffer, corresponding to the symbol's position in the symbol set. The IBM 3279 Model S3G can store up to six programmed symbol sets of 190 values each. Three of the symbol sets are "triple-plane," having 190 entries each for red, green, and blue. Triple-plane symbols allow for a choice of seven colors for every PEL. The use of programmed symbols is being rapidly replaced by vector graphics.

protected field A display field into which data may not be entered. The use of an alphanumeric key while the cursor is in a protected field will result in an input-inhibited condition.

record/play/pause A capability which allows keystrokes to be recorded in a special file and replayed on request. Used to avoid repetition of commonly used keystroke sequences.

reverse video Display attribute in which the foreground and background colors are exchanged (e.g., black character on green background, rather than green character on black background).

S3G Informal term for the IBM 3279 Model S3G, a now-obsolete terminal which provided seven-color support, an APL keyboard and programmed-symbol graphics. See *APL, programmed symbol,* and the *3279* entry in the *3270 Display Station* table in this section.

scan code Code transmitted by the keyboard each time a key is depressed or released. Each key's scan code is determined by its position, rather than by the character engraved, embossed, or printed on it. Translation of keystrokes or keystroke combinations is handled by the Cluster/Establishment Controller for Control Unit Terminals and, for Distributed Function Terminals, within the terminal itself.

screen format See *IBM 3278* in *Sec. 3*.

SCS SNA Character String. Control codes used to format a visual presentation medium, such as a printed page. LU Type 1 printers use SCS controls, whereas LU Type 3 printers print from a device buffer similar to that of a 3270 display terminal.

seven-color support Support for the colors red, blue, green, pink, turquoise, yellow, and white.

single-session Having access to only one host session at a time.

sixteen-color support Support on IBM 3270 graphics terminals and in 3270 graphics emulation products for the colors black, blue, dark blue, brown, gray, green, dark green, mustard, orange, pink, purple, red, turquoise, dark turquoise, yellow, and white. For the purpose of defining 16-color 3270 graphic support, the IBM 3270 PC/GX (no longer manufactured) is considered to be the standard.

status line See *OIA*.

structured field For both displays and printers, a special field, within the data stream, neither destined directly for nor originating from the presentation space. May be used to download programmed symbols and vector graphics and for the transfer of non-presentation-space data (e.g., for file transfer or the control of non-IBM peripheral devices such as plotters), etc. IBM's intelligent printer data stream (IPDS) makes use of structured fields.

symbol set A set of graphical display characters (e.g., letters, numbers, punctuation, other symbols) and the 3270 buffer code values associated with them. IBM 3270 symbol sets provide for a total of 190 values each and may be predefined (Symbol Set 0: basic 3270 set; Symbol Set 1: APL/Text) or may be loadable (see *programmed symbol*).

Terminal Multiplexer The IBM 3299, which can support up to 8 (Model 2 or 3) or up to 32 (Model 032) Category A devices. The 3299 may be connected via coaxial or twisted-pair cable to the terminals and to the terminal adapter of an IBM 3174, to a multiplexed port on an IBM 3274 (3299 Models 2 and 3 only) or to the multiplexed port of another 3299. The 3299 Model 032 also supports a fiber-optic connection to the 3174.

TokenWay 3174 Establishment Controller Low-cost version (Model 90R) of the IBM 3174 Establishment Controller, supporting one Category A terminal adapter, one 4- or 16-Mbit/s Token-Ring network, and an SNA/SDLC host communication link at up to 64 kbit/s. In conjunction with an IBM 3299 Terminal Multiplexer, the terminal adapter can support up to eight 3270 display terminals or printers.

type ahead Terminal (DFT mode) or cluster controller (CUT mode) function, allowing the entry and storage of keystrokes during a host input/output operation or a printer-busy condition.

type/value pair Pair of bytes in a 3270 data stream used to define a field attribute, extended field attribute, or character attribute. The type byte specifies field (for field attribute), color, extended highlighting, or symbol set. The value byte provides the field attribute value, the color number, the extended highlighting option, or the symbol set number. See also *Control Codes SA, SF* and *SFE,* under *EBCDIC,* in *Sec. 5*.

underscore Optional field or character attribute. One of three attributes displayable as part of Extended Highlighting (optional with some display terminals, standard with others).

unprotected field A field into which data can be entered.

vector graphics Graphics support characterized by the use of "drawing orders," which are used by the display terminal or graphics printer to create and display the desired graphics images. Drawing orders are not dependent on the geometry or other physical characteristics of the terminal device. With vector graphics, the display terminal operator can interact directly with the host graphics application to modify individual structures. This is in contrast to programmed-symbol graphics in which the image can only be displayed.

Vital Product Data Term referring to the identifying characteristics of a device. Consists of information on device type, model number, and serial number. IBM's network management product, NetView, transmits requests for Vital Product Data to physical units and uses the information to track network status and assets.

windowing Display technique in which multiple sessions or multiple application displays can be viewed concurrently within rectangular, bordered areas on the screen.

IBM 3270 Information Display System: Display Terminals

Machine	Model(s)	Screen Format	Scrn Size	Mono/ Color	CUT/ DFT	Keys	Prtr Port	Comments
3178 †	C10, C20 C30	24x80	12"	Green	CUT	75 or 87	No	Ergonomic version of basic 3278 Model 2. Models denote keyboard layout.
3179 †	1 (100)	24x80	14"	7 colors	CUT	122	No	Ergonomic successor to 3279 Model S2B and 02X. Seven-color support and extended highlighting are standard.
3179 †	G (G10, G20)	24x80 & 32x80	14"	8 colors	DFT	122	Yes	Graphics. Model G10 has typewriter layout. Model G20 has APL2/typewriter layout.
3180 †	1 (110, 120, 130)	24x80, 32x80, 43x80 & 27x132	15"	Green	CUT	122	No	APL keyboard available (Model 130). Supports vertical scrolling and keystroke record/playback.
3191 †	A, B (A10-A40, B10-B40)	24x80	12"	Green (A) Amber (B)	CUT	102, 104 or 122	No	Data stream compatible with 3278 Model 2 and 3178. Has no extended attribute buffer (i.e., no extended highlighting capability). See also *3191 D, E.*
3191	D, E (D10-D30, E10-E30)	24x80 & 32x80	14"	Green (D) Amber (E)	CUT	102, 104 or 122	Yes	For all 3191s and 3192s (except 3192 Model G), x10 models have 122 keys, x20 models have 104 keys, and x30 models have 102 keys. 3191 Models A40 and B40 have 104-key data-entry keyboards (U.S. English only).
3191 †	L (L10-L30)	24x80 & 32x80	14"	Green	CUT	102, 104 or 122	Yes	See *3191 D, E.*

Machine	Model(s)	Screen Format	Scrn Size	Mono/ Color	CUT/ DFT	Keys	Prtr Port	Comments
3192 †	C (C10-C30)	24x80 & 32x80	14"	7 colors	CUT	102, 104 or 122	Yes	See *3191 D, E.*
3192 †	D (D10-D30)	24x80, 32x80, 43x80 & 27x132	15"	Green	CUT	102, 104 or 122	Yes	This model replaced the 3180 Model 1. See also *3191 D,E.*
3192 †	F (F10-F30)	24x80, 32x80, 43x80 & 27x132	14"	7 colors	CUT	104 or 122	Yes	See *3191 D,E.*
3192 †	G (G10-G40)	24x80 & 32x80	14"	8 colors	DFT	104 or 122	Yes	Graphics. Models: G10—122-key typewriter; G20—122-key APL2/typewriter; G30—104-key typewriter; G40—104-key APL2/typewriter. Replaced 3179-G.
3192	L (L10-L30)	24x80 & 32x80	14"	7 colors	CUT	102, 104 or 122	Yes	See *3191 A, B.* Supports a lightpen.

† indicates machine no longer manufactured.
CUT = Control Unit Terminal (single host session, unless attached to 3174 port with MLT Level 6 support).
DFT = Distributed Function Terminal (if printer port = Yes, with one display session plus one optional host-addressable printer session). Graphics terminals may support a plotter.
Printer port on CUT-mode terminals is always for local-copy printing only.
Models with alphabetic second characters are not shown. IBM uses them to denote alternative warranties.

Machine	Model(s)	Screen Format	Scrn Size	Mono/ Color	CUT/ DFT	Keys	Prtr Port	Comments
3192 †	W (W10-W30)	24x80, 32x80, 43x80 & 27x132	15"	Black on white	CUT	102, 104 or 122	Yes	See *3191 A, B.*
3193	10, 20 (010, 020)	48x80, 24x80, 32x80, 43x80 & 27x132	15"	Black on white or white on black	DFT	122 (010) or 102 (020)	No	Image capability, with maximum image of 880x1200 PELs or a mixture of alphanumeric and image data. One or two logical screens and up to eight partitions, which may be scrollable or hidden.
3194	CD0-CF0, C10-C30, DD0-DF0, D10-D30, HD0-HF0, H10-H30 H50 †	24x80, 32x80[1,2], 43x80[2]& 27x132[2]	12", 14" & 15"	*(See comments)*	CUT or DFT	102, 104 or 122	Yes[3]	Cxx models are 12-in, 8-color; Dxx models are 15-in, mono; Hxx models are 14-in, 8-color. xD0 and x10 models are 122-key; xE0 and x20 models are 102-key; xF0 and x30 models are 104-key. x10, x20 and x30 models have a 1-year warranty; xD0, xE0 and xF0 models have a 3-year warranty. The 3194 is a "high-function" programmable terminal, with optional diskdrive and host download capability. A Device Function Interface provides compatibility with the 3270-PC HLLAPI (high-level language application programming interface). Operation is similar to that of the 3270-PC, with up to four host sessions, up to two notepads, and a "utility" session (equivalent to a 3270-PC DOS session). A VT100-compatible ASCII display session is also available via an optional serial port.

[1] Only DFT mode on C models. [2] Either CUT or DFT mode on D and H models; 27x132 not available on C models.
[3] IBM Proprinter supported as local printer only.

Machine	Model(s)	Screen Format	Scrn Size	Mono/ Color	CUT/ DFT	Keys	Prtr Port	Comments
3270 Personal Computer †		24x80, 32x80[1], 43x80[1]& 27x132[1]	12"	8 colors[2]	CUT or DFT	101[3], 102 or 122	Yes	Supports one host display session (CUT mode) or up to four host display sessions (DFT mode), plus up to two notepad sessions and one or more DOS sessions. Includes hardware-controlled windowing, with a Workstation Control Mode, allowing windows to be moved, sized, and colored. Versions include 3270 PC and 3270 PC-AT (text and optional programmed-symbol graphics); 3270 PC/G and 3270 PC-AT/G (text and vector graphics); and 3270 PC/GX and 3270 PC-AT/GX (text and high-resolution vector graphics, including manipulation of 3D images). APL2 support is available on all models (DFT mode only). The printer port can only be used for local-copy or DOS application printing.
		[1] Full screen viewable only by cursor scrolling		[2] GX models support 16 colors		[3] U.S. only		
3276 †	1-4		12"	Green	CUT	75 or 87	Co-axial	Combined cluster controller and display station, supporting 2, 4, 6, or 8 CUT-mode displays (including itself) and/or printers. BSC models (001-004) cannot support attached displays having a larger screen format or more comprehensive feature set than they do. BSC models communicate at up to 7200 bit/s. SDLC models (011-014) communicate at up to 9600 bit/s.
	001/011	12x80						
	002/012	24x80						
	003/013	32x80						
	004/014	43x80						

† indicates machine no longer manufactured.
CUT = Control Unit Terminal (single host session, unless attached to 3174 port with MLT Level 6 support)
DFT = Distributed Function Terminal (if printer port = Yes, with one display session plus one optional host-addressable printer session). Graphics terminals may support a plotter.
Printer port on CUT-mode terminals is always for local-copy printing only.
Models with alphabetic second characters are not shown. IBM uses them to denote alternative warranties.

Machine	Model(s)	Screen Format	Scrn Size	Mono/ Color	CUT/ DFT	Keys	Prtr Port	Comments
3278 †	1-5		12"	Green	CUT	75 or 87	No	This is the original Category A display terminal. Although it has a detachable keyboard, its design is not ergonomic.
	001	12x80						
	002	24x80						Options include extended highlighting, programmed symbols, APL, lightpen, and badge reader.
	003	32x80						
	004	43x80						
	005	24x80 & 27x132						
3279 †	2A-3B		12"		CUT	75 or 87	No	An A in the model number indicates base color mode (defined elsewhere in this Glossary). All other models support 7 colors (8 colors for S3G), extended highlighting, and APL.
	A02	24x80		Base				
	A03	32x80		Base				Programmed symbol graphics support is standard on the S3G and optional on the 03X.
	B02	24x80		7 colors				
	B03	32x80		7 colors				Standard models (with S in the model number) are not upgradable.
	S2A-3X							
	S2A	24x80		Base				
	S2B	24x80		7 colors				
	S3G	32x80		8 colors				
	02X	24x80		7 colors				
	03X	32x80		7 colors				

Machine	Model(s)	Screen Format	Scrn Size	Mono/ Color	CUT/ DFT	Keys	Prtr Port	Comments
3290 †	220, 230	62x160, 50x106	17"	Orange (plasma)	DFT	122	No	The 3290 was IBM's first distributed function terminal (DFT). It supports up to five logical sessions and can simultaneously display four emulated 3278 Model 2s (full screen). APL2, programmed symbols and Entry Assist are standard.
3471 InfoWindow		24x80	14"		CUT	102, 104 or 122	No	E models are data-stream-compatible with the 3191 Models A and B and are the current successors to the 3278 Model 2 (no extended data stream support—i.e., no extended highlighting). B models, introduced in February 1991, provide Extended Vital Product Data, extended data stream support, Record/Play/Pause with security options, Country Extended Code Page, Improved Setup Mode, and front-of-screen dot-width adjustment..
	BA1, BA3			Amber-gold				
	BG1, BG3			Green				
	EA1, EA3			Amber-gold				
	EG1, EG3			Green				
3472 Infowindow		24x80, 32x80, 43x80 & 27x132			DFT	102, 104 or 122	Yes	Successor to the 3192 Models D, F, and W. No lightpen support. (3192 Model L is still available and is IBM's "product of choice" for lightpens.) x= 1 for 1-year warranty; x = 3 for 3-year warranty. For Europe, Middle East, and Africa, IBM offers Models C10 to C30 (14-in color), D10 to D30 (15-in green), E10 to E30 (15-in amber-gold), and W10 to W30 (black on white). x10, x20, and x30 signify 122-key typewriter, 102-key (Enhanced), and 122-key APL/typewriter, respectively.
	FAx, HAx		14"	Amber-gold				
	FCx, GCx, MCx		14"	7-color				
	FDx, HDx		15"	Green				
	FGx, HGx		14"	Green				

† indicates machine no longer manufactured.
CUT = Control Unit Terminal (single host session, unless attached to 3174 port with MLT Level 6 support);
DFT = Distributed Function Terminal (if printer port = Yes, with one display session plus one optional host-addressable printer session). Graphics terminals may support a plotter. Printer port on CUT-mode terminals is always for local-copy printing only.
Models with alphabetic second characters are not shown. IBM uses them to denote alternative warranties.

Machine	Model(s)	Screen Format	Scrn Size	Mono/ Color	CUT/ DFT	Keys	Prtr Port	Comments
3472 InfoWindow Graphics-5			14"	8 colors	DFT-E	102, 104 or 122	Yes	Enhanced successor to the 3194 Models C, D, and H, plus the graphics capability of the 3192 Model G. For session-support capabilities, see the comments for these terminals.
	GCx, MCx	24x80, 32x80, 43x80 & 27x132						MCx models support a wider range of attached printers than GCx models. 122-key keyboard may be typewriter or APL/typewriter. DFT-E is an extension of DFT, allowing access to an ASCII host session, controller customization and access to X.21/X.25 and 3174 diagnostics. For Europe, Middle East and Africa, IBM offers models G10 to G30, which have 122-key typewriter, 102-key (Enhanced) and 122-key APL/typewriter keyboards, respectively.

CUT = Control Unit Terminal (single host session, unless attached to 3174 port with MLT Level 6 support)
DFT = Distributed Function Terminal (if printer port = Yes, with one display session plus one optional host-addressable printer session). Graphics terminals may support a plotter.
Printer port on CUT-mode terminals is always for local-copy printing only.

IBM 3270 Information Display System: Printers

Machine	Model(s)	Type	Color	Print speed	Char/ line	Line/ inch	Char set	Comments
3262	1, 3, 11, 13	Line	No	Line/min	132	3, 4, 6 or 8	48, 64, 96 or 128	
	001, 003			253-650				Print speed is inversely proportional to the size of the character set.
	011, 013			125-325				
3268 †	2, 2C	Serial matrix		340 char/s	132 or 220	3, 4, 6 or 8	94 or 222	
	002		No					
	C02		Yes					
3287 †	1,2,1C,2C	Serial matrix		Char/s	132	6,8	94 or 122	Color models support a base color mode, similar to that of the 3279 display terminal. Extended color printing is limited to the primary colors, with the secondary colors and white printed in black. Optional programmed symbol support is similar to that offered with display terminals such as the 3279 Model S3G. However, since the character matrix differs from that of the display terminals, the symbol set must be separately downloaded by the host application.
	001		No	80				
	002		No	120				
	C01		Yes	80				
	C02		Yes	120				
3812 Page-printer	2	Laser	No	Up to 12 page/min	V	V	V	Requires 3270 Attachment Feature #3190. Resolution is 240x240 dot/inch. Supports host graphics, font selection, etc., via the intelligent printer data stream (IPDS).

† indicates machine no longer manufactured.
For characters per line, lines per inch, and character set, V indicates variable. In some cases, a finite set is defined in the IBM specifications. However, in many cases, that set has too many values to list here.

Machine	Model(s)	Type	Color	Print speed	Char/ line	Line/ inch	Char set	Comments
3816 Page-printer	1S, 1D	Laser	No	Up to 24 page/min	V	V	V	Requires 3270 attachment feature #7653. In addition to the same IPDS capabilities as the 3812, also supports 3268 compatibility. Model 1S is simplex (single-sided printing); Model 1D is duplex. Duplex operation requires IPDS.
4224	201-2E3	Serial matrix		Char/s	V	3, 4, 6 or 8	V	IPDS and non-IPDS modes.
	201		No	200				
	202, 2E2		No	400				
	2C2		Yes	400				
	2E3		No	600				
4234	1, 011	Dot band	No	Line/min	V	3, 4, 6 or 8	V	Print speed depends on choice of draft, data processing, or near-letter-quality, and on number of characters and lines per inch. Model 011 operates in both IPDS and non-IPDS modes. Model 001 is non-IPDS only.
	001			60-410				
	011			115-800				
4245	D12, D20	Line band	No	Line/min	132	6 or 8	48, 50, 52, 63, 94, 98 or 124	
	D12			1200				
	D20			2000				
4250 †	1 (001)	Elec-tro-ero-sion	No	1.5 to 2.5 min/page	V	V	V	Produces typeset-quality output, using a dry process, on aluminum-coated roll paper. Resolution is 600 dot/inch.

Machine	Model(s)	Type	Color	Print speed	Char/ line	Line/ inch	Char set	Comments
4250 II † Electro-compositor	II (002)	Electro Erosion	No	1.5 to 2.5 min/page	V	V	V	Produces typeset-quality output, using a dry process, on aluminum-coated roll paper or polyester. 4250 Model 1 can be upgraded to this model.
4250 V Electro-compositor	V (005)	Electro Erosion	No	1.5 to 2.5 min/page (A4 or letter size)	V	V	V	Produces typeset-quality output, using a dry process, on aluminum-coated roll paper or polyester. Maximum paper width is 43.8 cm (17.25 in), which accommodates formats up to A2 size.
5210 †	G1, G2	Wheel	No	Char/s	132	3.4, 5.33, 6, 8, 9.6, 12, 24 or 48	96	
	G01			40				
	G02			60				

† indicates machine no longer manufactured.

For characters per line, lines per inch, and character set, V indicates variable. In some cases, a finite set is defined in the IBM specifications. However, in many cases, that set has too many values to list here.

IBM 3270 Information Display System: Noncoaxial Printer Attachment

Printer	3174 Establishment Controller[1]	3179 Model G	3191 Models D, E & L	3192 All Models	3192 Model G	3194 All Models[2]	3270-PC[3] All Models	3472 and 3472-G
3852-2 Color Jetprinter †		√			√		√	√[4]
4201-1 Proprinter †	√		√	√	√	√	√	
4201-2 Proprinter II	√		√	√	√	√	√	√
4201-3 Proprinter III	√		√	√	√	√		√
4202-1 Proprinter XL †	√		√	√	√	√	√	
4202-2 Proprinter II XL	√		√	√	√	√		√
4202-3 Proprinter III XL	√		√	√	√	√		√
4207-2 Proprinter X24E	√				√			√
4208-2 Proprinter XL24E	√				√			√
5201-1 Quietwriter							√	
5201-2 Quietwriter II							√	
5202-1 Quietwriter III							√	√
5204-1 Quickwriter	√						√	

† Indicates printer no longer manufactured.

[1] Requires Asynchronous Emulation Adapter feature on the 3174 and a serial interface on the printer.

[2] Except Model H50, which supports only the 4201 Model 1 Proprinter.

[3] Additional printers supported on the 3270-PC are the 3852-1 and 5182 Color Printers, 3812 Pageprinter, 5216 Wheelprinter, 5223 Wheelprinter E, and 5152 Graphics Printer.

[4] 3472-G Model MC only (which also supports HP ThinkJet, DeskJet, and PaintJet, and Epson FX1050 and LQ1050)

IBM 5250 Information Display System

The IBM 5250 Information Display System (5250 IDS) has, for several years, been the standard terminal subsystem for the IBM System/36, System 38, and, more recently, the Application System/400 (AS/400). It was developed by a different IBM Division from the one which developed the 3270 Information Display System. Not surprisingly, it exhibits considerable architectural differences and, in its documentation, a large number of differences in nomenclature.

IBM's more recent emphasis on PC-based products, such as AS/400 PC Support, incorporating Logical Unit Type 6.2 (APPC capability), has reduced the emphasis on the 5250 IDS. Much of IBM's own AS/400-based application software has been upgraded accordingly. However, there are many thousands of proprietary applications which continue to support the 5250 data stream. Thus, for many, an understanding of the architecture and features and a knowledge of the devices of the 5250 Information Display System continues to be important.

IBM 5250 Information Display System Terminology

The following definitions can provide a general understanding of the operation of the display terminals of the IBM 5250 IDS. Although they are useful as a reference for the 5250 data stream programmer, they do not provide all the information required for the programming task. Those requiring a fuller understanding might want to start with the *IBM 5250 Information Display System Functions Reference Manual* (SA21-9247).

active field A 5250 display terminal field into which the operator has begun to enter data.

attribute byte A byte, within the display buffer, containing the screen attribute information for the bytes following it. For monochrome displays, there are five possible display attributes, and they can appear in any combination that does not include all three of underscore, high intensity, and reverse image.

Bit Definitions in 5250 Screen Attribute Byte

Bits	Value	Meaning
0-2	001	Identifies the byte, uniquely, as a screen attribute byte
5-7	111	Non-display *(See below for other values of these bits)*
		Monochrome Displays
3	0 1	No column separator Column separator
4	0 1	No blink Blink
5	0 1	No underscore Underscore
6	0 1	Normal intensity High intensity
7	0 1	Normal image Reverse image

For color displays, there is no simple way of combining a set of binary options, and certain combinations are simply not available. The following table shows the possible combinations.

Available Combinations of Emphasis and Color for 5250 Color Displays

	Color						
	Blue	Green	Turq.	Red	Pink	Yellow	White
Emphasis	Values below are for bits 3 to 7 of screen attribute byte						
None	11010	00000		01000	11000		00010
Underscore only	11110	00100		01100	11100		00110
Blink only				01010			
Underscore and blink				01110			
Reverse image only	11011	00001		01001	11001		00011
Underscore and rev. image		00101		01101	11101		
Reverse image and blink				01011			
Column separators only			10000			10010	
Underscore and column sep.			10100			10110	
Reverse image and col. sep.			10001			10011	
Underscore, rev. im., and col. sep.			10101				

The screen attribute byte occupies a buffer position and hence a screen position, immediately to the left of the field to which it applies. It displays as a space. See also *Field Format Word*.

Cable-Thru Feature of 5250 display and printer terminals, in which two connectors are provided, allowing end-to-end connection of terminals on the same controller port. Cable-Thru may be used with twinaxial or twisted pair cable, but they may not be mixed.

error line Row on which error messages are displayed. Error messages are preceded, in the 5250 data stream, by a Write Error Code command (hex 21), an Insert Cursor command (hex 13), and a cursor address (row and column). The error line can be specified in the format table header or may be allowed to default to row 1.

error state

hardware The machine state that exists when a hardware malfunction is detected. No keystrokes can be processed. Clearing the hardware error state puts the display in the normal unlocked state, with Format level 0 selected.

posthelp State that may be entered, by the depression of the Help key, only if an error has occurred. A signal is sent to the host, allowing the host application to transmit an appropriate help screen.

prehelp The state that exists when the operator makes a keying error or when the incoming data stream contains the Write Error Code command (see error line). The Reset key terminates this state. The help key puts the display in the posthelp error state.

Field Control Word (FCW)) A 2-byte optional field, associated with and immediately following, in the format table, a Field Format Word. There are four types of Field Control Words, specifically:

- Magnetic stripe reader
- Resequencing
- Selector lightpen
- Self-check (modulo 10 or modulo 11)

The magnetic stripe reader and selector lightpen functions can also be combined. In recent terminals, the magnetic stripe reader function has been expanded to include bar-code wands and bar-code slot readers. The following table summarizes the Field Control Word options.

Bit Definitions in 5250 Field Control Word

Byte 1 Binary (Hex)	Byte 2 Binary (Hex)	Type and Description
10000000 (80)	00000000 (00) to 11111111 (FF)	**Resequencing.** The value of the second byte corresponds to the ordinal position, on the screen, of the associated field. Fields are sent to the host in the order of occurrence of the associated FCWs in the format table. The last FCW must have the value 11111111 (FF) in the second byte in order to stop the transfer. If the first FCW has a second byte value of all zeros, fields are sent to the host in their order of occurrence on the screen.
10000001 (81)	00000001 (01)	**Magnetic stripe reader with operator ID secure data.** (If operator ID secure data is not to be entered, a Field Control Word is not required.)
	00000010 (02)	**Selector lightpen**.
	00000011 (03)	**Both magnetic stripe reader with operator ID secure data and selector lightpen.**
10110001 (B1)		**Self-checking** (Self-check feature must be installed.) Field may be a maximum of 33 digits (including sign if signed numeric). Checking is based on the value of the 4 low-order bits of each entered character. Values of 0000 (0) to 1001 (9) are used unchanged; zero is substituted for values of 1010 (A) to 1111 (F).
	10100000 (A0)	Modulo 10.
	01000000 (40)	Modulo 11.

Field Format Word (FFW) A 2-byte field associated with a 5250 display field but not occupying a position in the display buffer. The ordinal position, in the field table, of the Field Format Word corresponds to the ordinal position, on the display, of the field which it qualifies.

Bit Definitions in 5250 Field Format Word

Byte 1: 0	1	2	3	4	5	6	7	Byte 2: 0	1	2	3	4	5	6	7
0	1	By-pass	DUP	MDT	Shift/Edit			Auto Enter	Field Exit	Mono Case		Mand-atory Enter	Right-Adjust/-Mandatory Fill		

Byte, bit(s)	Description
1, 0-1	Set to 01, to identify the Field Format Word
1, 2	Bypass. An attempt to enter anything will cause an error (same as 3270 protected field). 0 = Not a bypass field. 1 = Bypass field.
1, 3	Duplication (application-dependent function) 0 = DUP key not enabled for field. 1 = DUP key enabled for field, which will fill field with DUP codes (*).
1, 4	Modified Data Tag 0 = Field not modified. 1 = Field has been modified (or host has arbitrarily set).

Byte, bit(s)	Description
1, 5-7	Shift/Edit Specifications 000 = Alphabetic shift (accepts all characters). 001 = Alphabetic only (upper- and lowercase letters and space). 010 = Numeric shift (accepts all characters) 011 = Numeric only (accepts only 0-9, +, –, comma, period, and blank). Field Exit, Field+ or Field– must be used to exit. Field– changes high-order 4 bits of final digit to D (hex), unless final digit is not 0-9, in which case an error results. 100 = Katakana (or other second language) shift. 101 = Digits only (accepts only 0 to 9 or, if DUP enabled, DUP key).
	110 = Auxiliary input only (magnetic stripe reader, wand, selector lightpen —see Field Control Word). 111 = SIgned numeric (accepts only 0-9). Field Exit, Field+, or Field– must be used to exit. Field– changes high-order four bits of final digit to D (hex).
2, 0	Auto Enter 0 = No Auto Enter. 1 = Auto Enter—When the last character is entered, or when one of the Field Exit keys is used, all fields with MDT (byte 1, bit 4) set to 1 will be sent to the host.
2, 1	Field Exit Required 0 = Field Exit key is not required. 1 = The operator can exit the field only with a Field Exit key.
2, 2	Monocase 0 = Accept lowercase letters as entered. 1 = Convert lowercase letters to uppercase.
2, 4	Mandatory Enter 0 = Not Mandatory Enter. 1 = Operator must enter something into this field (and all similar fields) in order for the Enter key (or Auto Enter field) to be active.
2, 5-7	Right Adjust/Mandatory Fill (001 to 110 are not used) 000 = No Right Adjust or Mandatory Fill. 101 = Right Adjust with Zero Fill—Characters are right-adjusted when the field is exited and zeros are placed in unoccupied leading positions. 110 = Right Adjust with Blank Fill—Characters are right-adjusted when the field is exited and spaces are placed in unoccupied leading positions. 111= Mandatory Fill—The operator must fill this field in order to exit from it.

Format level 0 The condition of the format table at power-on or reset time. It contains one Field Format Word, containing all zeros (except for the 01 in bits 0 and 1 of byte 1). In the screen buffer, the associated single field has its attribute byte at row 1 column 1 and extends from row 1 column 2 to column 80 of the last row. The error line is row 1.

format table A table, associated with a 5250 display station, containing header information (optional), Field Format Words, and, where applicable, Field Control Words. The table's maximum capacity is 127 words (254 bytes). In the absence of header information and Field Control Words, the characteristics of up to 127 fields may be specified.

format table header Seven-byte optional field at the beginning of the format table. The following table describes the function of each of the seven bytes.

Function Table Header Bytes

Byte	Description
1	Reserved.
2	Format ID — 00 to FF (hex).
3	Resequencing. 00 = no resequencing (ignore resequencing Field Control Words, if any). *nn* = resequencing, starting with the field associated with the *nn*th Field Control Word. (See *Field Control Word.*)
4	Row for error line. If 0 or out of range, error line defaults to last row of display.
5-7	Include data mask for Command Function keys. The bits from byte 5, bit 0 to byte 7, bit 7 correspond to Command Function keys 24 to 1, respectively. For each position, a 1 indicates that all fields with the MDT bit set to 1 are to be sent to the host when the correponding Command Function key is depressed (comparable to 3270 PF key). A 0 indicates that the host is to be informed only of the depression of the corresponding Command Function key (comparable to 3270 PA key).

IPDS Defined under *IBM 3270 Information Display System Terminology,* at the beginning of this section.

keyboard/display terminal Terminal capable of supporting one or more operator to host keyboard/display sessions.

keyboard state

normal locked state The state that exists after the operator has pressed a key requiring host attention or as a result of one of a number of host commands.

normal unlocked state The normal state of the keyboard, in which keystrokes are accepted, key click is enabled, the input inhibited indicator is off, and invalid keys will cause errors. This state is entered as a result of a host command or of the depression of the Error Reset key.

SS message state The state that is entered as a result of a command sent by Supervisory Services on the host system. The keyboard is locked and the input inhibited indicator is on. Cleared by the depression of the Error Reset key.

System Request state The state that is entered by the depression of the Sys/Req (System Request) key. This is the state in which communication with Supervisory Services on the host system takes place. This state can be entered directly from the normal locked state.

mandatory enter 5250 display terminal field type requiring that the operator enter something in the associated field before proceeding. Where a formatted display includes mandatory enter fields, none of them can be bypassed.

mandatory fill 5250 display terminal field type requiring that, once any data has been entered, the associated field be filled.

message line Row on display screen reserved for the display of operator error information. Treated, by the host, as a separate logical unit (LU Type 4) from the display as a whole (LU Type 7) and addressed via an SS-LU (Supervisory Services to logical unit) session. Not available on older 5250 displays.

mode Any of the following three modes can be entered when the display terminal is in the normal unlocked keyboard state:

command Mode that is entered as a result of the depression of the Cmd key. Not all 5250 displays have a Cmd key. Allows the use of one of 24 preprogrammed Command Function keys, whose meaning is application-dependent. This mode ends when a Command Function key or the Error Reset key is depressed.

data Mode in which normal entry of data takes place.

insert Mode that is entered as a result of the depression of the Ins key. The insert indicator is displayed and alphanumeric keystrokes may be used to insert data ahead of existing data, subject to the constraints imposed by Field Format Words. This mode ends with a return to data mode when the Error Reset key is depressed.

operator error line Display screen row, specified in the format table header, on which operator errors are to be displayed. If the format table header specifies no error line row number or specifies an out-of-range number, the operator error line defaults to the message line, if supported.

power-on state The state, immediately after power on, in which all keystrokes not requiring host action can be accepted (subject to the controller being powered on). Format level 0 (see also) is active.

record/play/pause A capability which allows keystrokes to be recorded and replayed on request. Used to avoid repetition of commonly used keystroke sequences.

reverse image Display attribute in which the foreground and background colors are exchanged (e.g., black character on green background rather than green character on black background).

screen attribute Any of several characteristics of a field (one or more successive characters). For a detailed description, see *attribute byte*.

screen format The number of displayable rows and columns. Available 5250 screen formats are 24x80, 27x132, 24x80+17x80 (split-screen), and, for newer terminals (e.g., 3477), 24x80+18x80. Newer terminals also have an additional row—the status line or Operator Information Area (OIA). IBM's most recent terminals (3487 and 3488 InfoWindow II) also support 32x80, 43x80, and variable split screen.

seven-color support Support for the colors red, blue, green, pink, turquoise, yellow, and white.

single-session Having access to only one host session at a time.

twinax Abbreviation for twinaxial cable.

twinaxial cable A cable similar to coaxial cable, except that there are two center conductors rather than one. Twinaxial cable can support multiple connections along its length, using T-connectors (not required with IBM terminals having the "Cable-Thru" feature) at the intermediate connection points. The twinaxial cable used for IBM 5250 series terminals has a larger diameter than the coaxial cable used with IBM 3270 series terminals.

twisted-pair cable Optional substitute for twinaxial cable (see general definition in *Sec. 1*). Can be connected directly to the newer terminals, host workstation adapters, and the Twinax to Telephone Twisted Pair Adapter (TTPA) of the IBM 5299 Terminal Multiconnector. Can connect to older twinaxial devices via individual TTPAs.

type ahead Terminal function, allowing the entry and storage of keystrokes (typically up to 32) during a host input/output operation.

underscore Optional screen attribute.

IBM 5250 Information Display System: Display Terminals

Machine	Model(s)	Screen Format	Scrn Size	Mono/Color	Keys	Prtr Port	Comments
3179 †	2	24x80	14"	7 colors		No	
	200				122		
	220				102		
3180 †	2 (210 & 220)	24x80 & 27x132	15"	Green	122	No	
3196 †	A, B	24x80	12"			No	
	A10			Green	122		
	B10			Amber-gold	102		
	A20			Green	122		
	B20			Amber-gold	102		
3197 †	C	24x80	14"	7 colors		Yes	The middle digit of the model number indicates the warranty period (numeric = 1 year, alphabetic = 3 years). Printer is host-addressable but cannot be addressed if display is in split-screen mode (24x80+18x80). Any of the printers (except the 3852-2) in the table on page 222 may be attached.
	C10, CD0				122		
	C20, CE0				102		
	D	24x80, 27x132 & 24x80 + 17x80	15"	Green			
	D10, DD0				122		
	D20, DE0				102		
	D40, DG0				122[1]		
	W			Black on white			
	W10, WD0				122		
	W20, WE0				102		

† indicates machine no longer manufactured

[1] Data entry, U.S. English only.

Machine	Model(s)	Screen Format	Scrn Size	Mono/Color	Keys	Prtr Port	Comments
3476 InfoWindow		24x80	14"		122 (type-writer or data entry)	No	The last digit of the model number indicates the warranty period, in years. B models, introduced in March 1991, have front-of-screen dot-width adjustment.
	BA1, BA3			Amber-gold			
	BG1, BG3			Green			
	EA1, EA3			Amber-gold			
	EG1, EG3			Green			
3477 InfoWindow		24x80, 27x132 & 24x80 + 18x80			122 or 102	Yes	Printer is host-addressable but cannot be addressed if display is in split-screen mode (24x80+18x80). Any of the printers (except the 3852-2) in the table on page 222 may be attached. The last digit of the model number indicates the warranty period, in years.
	FA1, FA3		14"	Amber-gold			
	FC1, FC3		14"	7 colors			
	FD1, FD3		15"	Green			
	FG1, FG3		14"	Green			
3486 InfoWindow II		24x80	14"		122 or 102	Yes	122-key keyboard may be typewriter or data entry layout. Supports variable split-screen (horizontal or vertical) and DDP* (display-display-printer) modes of operation. The last digit of the model number indicates the warranty period, in years.
	BA1, BA3			Amber-gold			
	BG1, BG3			Green			
3487 InfoWindow II		24x80, 32x80, 43x80 & 27x132			122 or 102	Yes	122-key keyboard may be typewriter or data entry layout. Supports variable split-screen (horizontal or vertical) and DDP* (display-display-printer) modes of operation. Allows downloading of printer definition tables for support of a variety of ASCII printers. Has on-screen calculator capability. The last digit of the model number indicates the warranty period, in years.
	HA1, HA3		15"	Amber-gold			
	HC1, HC3		14"	7 colors			
	HG1, HG3		15"	Green			

† indicates machine no longer manufactured.
* DDP mode supports two independent display sessions and a concurrent, independent printer session.

Machine	Model(s)	Screen Format	Scrn Size	Mono/Color	Keys	Prtr Port	Comments
3488 InfoWindow II		24x80, 32x80, 43x80 & 27x132	N/A	Can use any VGA or VGA-compatible display	122 or 102	Yes	Has the same characteristics as the 3487 InfoWindow II, except that the user may choose any VGA (e.g., IBM 8504, 8511, 8512, 8514, 8515, 8518, 9515, or 9518) or VGA-compatible monitor.
	H11, H13						
5251 †	1, 11, 999		12"	Green	75	No	This is the original 5250 twinaxial display station. Although it has a separate keyboard, its design is not ergonomic.
	001	12x80					
	011, 999	24x80					
5251 †	2, 12		12"	Green	75	No	Combined SDLC controller and display terminal. Supports up to 8 additional 5250 or compatible displays and/or printers.
	002	12x80					
	012	24x80					
5252 †	1 (001)	12x80	12"	Green	75	No	Dual display (back to back), with two separate keyboards.
5291 †	1, 2	24x80	12"	Green		No	
5292 †	1,2	24x80	12"				
	001			7 colors		No	
	002			8 colors (graphic)			

† indicates machine no longer manufactured

IBM 5250 Information Display System: Printers

Machine	Model(s)	Type	Color	Print Speed	Char/ line	Line/ inch	Char Set	Comments
3812	2 (002)	Laser	No	Up to 12 page/min	V	V	V	V = variable.
4214 †	2 (002)	Serial matrix	No	Char/s	132 to 220	3, 4, 6 or 8	192 (EBCDIC)	
				50 (NLQ)				
				200 (DP)				
5219 †	D1, D2	Daisy wheel	No	Char/s	Up to 172 (at 15 per inch)	5.33, 6 or 8	96	
	D01			40				
	D02			60				
5224 †	1, 2	Line matrix	No	Line/min	132 to 198	6 or 8		Print speed is inversely proportional to horizontal print density (in char/inch).
	001			95-140				
	002			170-240				
5225 †	1, 2, 3, 4	Line matrix	No	Line/min	132 to 198	6 or 8		Print speed is inversely proportional to horizontal print density (in char/inch).
	001			195-280				
	002			290-400				
	003			355-490				
	004			420-560				

† indicates machine no longer manufactured

Machine	Model(s)	Type	Color	Print Speed	Char/ line	Line/ inch	Char Set	Comments
5256 †	1, 2, 3	Serial matrix	No	Char/s	132	6 or 8	95 or 184	
	001			40				
	002			80				
	003			120				
5262	1 (001)	Line band	No	130 to 650 line/min	132	6 or 8	48, 63, 64, 96 or 188	Print speed is inversely proportional to the size of the character set.

† indicates machine no longer manufactured

Section

5

Interchange Codes

Introduction

Of the several interchange codes currently in use, only two have found acceptance as general-purpose codes.

Extended Binary Coded Decimal Interchange Code (EBCDIC) is the de facto standard for IBM and compatible mainframe computers and for many communications devices associated with them. IBM introduced EBCDIC in 1964 with the announcement of System/360. Prior to the System/360, IBM offered several computer families, each of which had its own internal data structure [6-bit binary-coded decimal (BCD) characters for the 1401, 1440, 1410, 7010, and 7080, 10-decimal-digit words for the 7070, 7072, and 7074, and 36-bit words for the 7040, 7044, 7090, and 7094]. Data communication between computer and computer or between computer and terminals used a variety of codes, including four-of-eight code and BCDIC (binary-coded decimal interchange code), all of which were different from the computers' internal codes. With EBCDIC, IBM offered to create order out of chaos. Along with IBM's uniform mainframe architecture and a number of reasonably uniform protocols, EBCDIC also helped to create a market for IBM plug-compatible devices and subsystems, including disk drives, tape drives, terminals, and, ultimately, mainframe computers themselves.

Although a de facto standard, EBCDIC has fallen far short of achieving a monopoly position. ASCII (and its international variants), once presented as an alternative standard to EBCDIC, has enjoyed widespread acceptance, especially in the asynchronous (start-stop) video terminal marketplace. IBM, which finally joined the world of ASCII with its introduction in 1980 of the 3101 ASCII

Display Terminal, adopted, not EBCDIC, but an 8-bit extension of ASCII for its Personal Computer.

The Personal Computer's Extended ASCII has, like EBCDIC, become a de facto standard. With its fairly recent code-page support, IBM has produced a number of international variants of PC Extended ASCII. In the process, it has established a considerable stake in a standard it once rejected.

This section deals, not with ASCII specifically, but with the well-established 7-bit international standard known both as CCITT International Alphabet No. 5 and as ISO 646.

CCITT International Alphabet No. 5 (ISO 646)

This interchange code is defined by CCITT Recommendation T.50 and by ISO 646:1983. Its U.S. implementation is defined by the ANSI X3.4 standard, which is usually known as American National Standard Code for Information Interchange (ASCII).

In the following table, the column headings provide the three high-order bits and the row headings provide the four low-order bits of the code point value.

ISO 646 (CCITT International Alphabet No. 5) Table
International Reference Version

Hex		0	1	2	3	4	5	6	7
	Binary	000	001	010	011	100	101	110	111
0	0000	NUL	DLE TC7	SPACE	0	@ ❸	P	❶	p
1	0001	SOH TC1	DC1 See note	!	1	A	Q	a	q
2	0010	STX TC2	DC2 See note	” ❶	2	B	R	b	r
3	0011	ETX TC3	DC3 See note	£ # ❷	3	C	S	c	s
4	0100	EOT TC4	DC4 See note	$ ¤ ❷	4	D	T	d	t
5	0101	ENQ TC5	NAK TC8	%	5	E	U	e	u
6	0110	ACK TC6	SYN TC9	&	6	F	V	f	v
7	0111	BEL	ETB TC10	’ ❶	7	G	W	g	w
8	1000	BS FE0	CAN	(	8	H	X	h	x
9	1001	HT FE1	EM	)	9	I	Y	i	y
A	1010	LF FE2	SUB	*	:	J	Z	j	z
B	1011	VT FE3	ESC	+	;	K	[❸	k	{ ❸
C	1100	FF FE4	FS IS4	, ❶	<	L	\ ❸	l	| ❸
D	1101	CR FE5	GS IS3	-	=	M	] ❸	m	} ❸
E	1110	SO	RS IS2	.	>	N	^ ❶	n	‾ ❶
F	1111	SI	US IS1	/	?	O	_	o	DEL

The characters in the middle of each of columns 2 to 7 (i.e., all characters except £ and $) correspond to the international reference version of ISO 646. The U.S. and U.K. versions differ only slightly from the international reference version. The U.S. version (ANSI X3.4) defines code points 23, 24 and 7E as #, $, and ~, respectively. The U.K. version (British Standard Data Code) defines code points 23, 24, and 5E as £, $, and ↑, respectively.

Explanation of Table Notes

TC1 to TC10 Transmission Control characters. These are intended to control or facilitate the transmission of information over communication links. ISO 646 does not describe how they are used.

FE0 to FE5 Format Effectors. These are used to control the layout and positioning of information on printed pages or display devices. FE2 to FE5 are intended for equipment for which the horizontal and vertical movements are effected separately. For a New Line (NL) function (combined CR and LF), FE2 (normally Line Feed) is used.

DC1 to DC4 Device Control characters. These are used to control local or remote ancillary devices. With the exception of the X-Off and X-On functions described on page 229, they are not used for transmission control.

IS1 to IS4 Interchange Separators. These are used to separate and qualify data logically. They need not be hierarchical. However, when they are used hierarchically, they are used in ascending order from IS1 to IS4 (US, RS, GS, FS).

❶ Some of these code points (22, 27, 2C, 5E, 60, and 7E) may be used for national-language implementations requiring diacritical marks (accents, etc.). ISO 646 defines two possible sequences for this purpose, namely *<character>BS<mark>* and *<mark>BS<character>*. The standard also allows for diacritical mark codes to have their normal meaning when not adjacent to a backspace. The possible substitutions are:

- Dieresis (also called umlaut or tremma) for the double quote (22). Used in forming ä, Ä, ë, ï, ö, Ö, ü, Ü, ÿ, and Ÿ.
- Acute accent for the apostrophe (27). Used in forming á, Á, é, É, í, Í, ó, Ó, ú, and Ú. In the ANSI X3.4 (ASCII) implementation, this code point usually serves both as an apostrophe and as a closing single quotation mark.
- Cedilla for the comma (2C). Used in forming ç and Ç.

- Circumflex accent for the caret (5E). Used in forming â, Â, ê, Ê, î, Î, ô, Ô, û, and Û.
- Grave accent for code point 60. Used in forming à, À, è, È, ì, Ì, ò, Ò, ù, and Ù. In the ANSI X3.4 (ASCII) implementation, this code point is used for an opening single quotation mark.
- Tilde for the overline (7E). Used in forming ã, Ã, ñ, Ñ, õ, and Õ. The ANSI X3.4 (ASCII) implementation uses this code point for a tilde character (not for use as a diacritical mark).
- Other possible diacritical marks, such as breves, carons, dots, double acute accents, macrons, ogoneks, and rings for any of the six noted code points, for support of languages such as Albanian, Czech, Hungarian, Polish, Romanian, Serbo-Croatian, and Turkish.

Code points 5E, 60, and 7E may also be used for other graphical characters if the seven defined national use characters (see ❸) are insufficient.

With the introduction of 8-bit extensions to ISO 646 (IBM PC Extended ASCII, Adobe PostScript, HP LaserJet variations, etc.), all of which include characters with which diacritical marks are already associated, the need for the type of approach defined in ISO 646 is considerably reduced.

❷ These two code points are intended for currency symbols. Code point 23 is used for # in the ANSI X3.4 implementation, and £ in several other implementations, including British Standard Data Code (BSDC). Code point 24 is used for $ in many implementations, including ANSI X3.4 and BSDC. Other implementations use the generic currency symbol, ¤, or a national currency symbol (e.g. ¢, ¥, *f*). For international communication, the meaning of code points 23 and 24 is a matter for agreement between the communicating parties.

❸ These seven code points (40, 5B, 5C, 5D, 7B, 7C, and 7D) are reserved for national use. The ANSI X3.4 implementation (which is also the ISO 646 reference version) is shown. Possible uses are for frequently occurring accented characters (such as é in French; ä, Ä, ö, Ö, ü, and Ü in German; and ñ and Ñ in Spanish), accented characters treated as separate letters (such as the Swedish å and Å), diphthongs (such as the Danish æ and Æ), other Roman variants [such as the Danish ø and Ø, the German ß (ess-tset), the Turkish undotted i, and the Icelandic eth and thorn characters] and special punctuation (such as the Spanish ¡ and ¿).

CCITT Recommendations T.51, Coded Character Sets for Telematic Services, and T.61, Character Repertoire and Coded Character Sets for International Teletex Service, define very broad extensions (both 7- and 8-bit) to International Alphabet No. 5.

Explanation of the Control Codes in Columns 0, 1, and 7 of the ISO 646 Table

ACK (06): Acknowledgment Used to indicate the satisfactory receipt of a message (block).

BEL (07): Bell Causes the receiving device to sound an audible signal.

BS (08): Backspace Moves the active position backward by one character position on the same line. On some devices, may cause movement from the first position of one line to the last position of the previous line or from the first position to the last position in the presentation space.

CAN (18): Cancel Indicates that preceding data is in error and should be ignored. The starting point of the data to be ignored is application- or device-dependent.

CR (0D): Carriage Return Moves the active position to the first position on the same line.

DC1-DC4 (11-14): Device Control Codes DC1 and DC2 are intended for turning on a device or a function. DC3 and DC4 are intended for turning off a device or a function. The exact meanings are usually equipment-dependent. However, DC1 and DC3 are used as X-On (transmitter on) and X-Off (transmitter off) codes, respectively, for flow control in asynchronous (start-stop) mode on a full-duplex communication link.

DEL (7F): Delete Used in paper-tape application to obliterate characters that have been punched in error. Also used as a media-fill or time-fill character. Normally deleted on arrival at a receiving device or application.

DLE (10): Data Link Escape Used to change the meaning of a specific number (usually 1) of following characters, usually for control purposes.

EM (19): End of Medium Identifies either the physical end of a medium or the end of a used portion of a medium or the end of the wanted portion of data recorded on a medium.

ENQ (05): Enquiry Used to request a response from the remote station on a point-to-point communication link. The first use of ENQ after a connection is established normally means "Who are you?" Subsequent use may or may not have the same meaning.

EOT (04): End of Transmission Indicates the conclusion of a transmission, consisting of one or more messages.

ESC (1B): Escape Used to indicate a special meaning or purpose for one or more subsequent characters.

ETB (17): End of Transmission Block Used to indicate the end of a block containing text that is to be continued in the next block.

ETX (03): End of Text Used to signify the end of a text or message.

FF (0C): Form Feed Advances the active position to the same character position on a predetermined line on the next form or page.

FS (1C): File Separator Used to indicate a logical boundary between files.

GS (14): Group Separator Used to indicate a logical boundary between items called groups (e.g., groups of records within a file).

HT (09): Horizontal Tabulation Advances the active position to the next predetermined character position on the same line.

LF (0A): Line Feed Advances the active position to the same character position on the next line. On some devices, may cause movement from the last line to the first line of the presentation space.

NAK (15): Negative Acknowledgment Used as a negative response, usually to indicate the need to retransmit a message (block).

NUL (00): Null Used for media-fill (usually) or time-fill. Does not affect the information content but may affect layout.

RS (1E): Record Separator Used to indicate a logical boundary between records.

SI (0F): Shift In Used to indicate that subsequent character codes are to represent the graphic characters of the standard character set.

SO (0E): Shift Out Used to indicate that subsequent character codes are to represent a new set of graphic characters defined as an extension to the standard character set. Where there are two or more additional character sets, a preceding escape sequence (using the ESC character) is used to specify which set is required.

SOH (01): Start of Heading Indicates that a message heading follows.

STX (02): Start of Text Indicates that text information follows.

SUB (1A): Substitute character Used to replace a character that is known to be in error. Intended to be inserted by automatic means.

SYN (16): Synchronous idle Used, in synchronous mode, to establish synchronization by character (i.e., to define character boundaries).

US (1F): Unit Separator Used to indicate a logical boundary between units of information within a record or other larger item.

VT (0B): Vertical Tabulation Advances the active position to the same character position on a predetermined line.

EBCDIC (Extended Binary Coded Decimal Interchange Code)

EBCDIC is the interchange code IBM introduced when it announced System/360 in 1964. There are many national-language implementations for both Roman and non-Roman alphabets. The table below shows the U.S. English implementation.

The table column headings provide the four high-order bits (0 to 3) and the row headings provide the four low-order bits (4 to 7) of the code point value. (IBM bit numbering, within a byte, is 0 to 7 from most to least significant.) The shaded characters are the normal minimum character set. The unshaded characters cannot be displayed or printed on all devices.

EBCDIC Table
U.S. English Implementation

Hex	0	0	1	2	3	4	5	6	7	8	9	A	B	C	D	E	F
	0,1 -	00				01				10				11			
1	2,3 -	00	01	10	11	00	01	10	11	00	01	10	11	00	01	10	11
0	0000	NUL	DLE	DS		SP	&	-	ø	Ø	°	µ	^	{	}	\	0
1	0001	SOH	DC1/SBA	SOS		Req SP	é	/	É	a	j	~	£	A	J	Num SP	1
2	0010	STX	DC2/EUA	FS	SYN	â	ê	Â	Ê	b	k	s	¥	B	K	S	2
3	0011	ETX	DC3/IC	WUS	IR	ä	ë	Ä	Ë	c	l	t	₧	C	L	T	3
4	0100	SEL	RES/ENP	BYP/INP	PP	à	è	À	È	d	m	u	ƒ	D	M	U	4
5	0101	HT/PT	NL	LF	TRN	á	í	Á	Í	e	n	v	§	E	N	V	5
6	0110	RNL	BS	ETB	NBS	ã	î	Ã	Î	f	o	w	¶	F	O	W	6
7	0111	DEL	POC	ESC	EOT	å	ï	Å	Ï	g	p	x	¼	G	P	X	7
8	1000	GE	CAN	SA	SBS	ç	ì	Ç	Ì	h	q	y	½	H	Q	Y	8
9	1001	SPS	EM	SFE	IT	ñ	ß	Ñ	`	i	r	z	¾	I	R	Z	9
A	1010	RPT	UBS	SM/SW	RFF	¢	!	¦	:	«	ª	¡	[	-	|	²	³
B	1011	VT	CU1	CSP/FMT	CU3	.	$	,	#	»	º	¿	]	ô	û	Ô	Û
C	1100	FF	IFS/DUP	MF	DC4/RA	<	*	%	@	ð	æ	Đ	‾	ö	ü	Ö	Ü
D	1101	CR	IGS/SF	ENQ	NAK	(	)	_	'	ý	¸	Ý	˝	ò	ù	Ò	Ù
E	1110	SO	IRS/FM	ACK		+	;	>	=	þ	Æ	þ	´	ó	ú	Ó	Ú
F	1111	SI	IUS/ITB	BEL	SUB	|	¬	?	"	±	¤	®	‗	õ	ÿ	Õ	EO
	4-7																

Explanation of the Control Codes in Columns 0, 1, 2, 3, and F of the EBCDIC Table

ACK (2E): Acknowledgment Used to indicate the satisfactory receipt of a block (frame).

BEL (2F): Bell Causes the receiving device to sound an audible signal.

BS (16): Backspace Moves the active position backward by one character position on the same line. On some devices, may cause movement from the first position of one line to the last position of the previous line or from the first position to the last position in the presentation space.

BYP (24): Bypass Same code point as *INP*.

CAN (18): Cancel Indicates that preceding data is in error and should be ignored. The starting point of the data to be ignored is application- or device-dependent.

CR (0D): Carriage Return Moves the active position to the first position on the same line.

CSP (2B): Control Sequence Prefix Same code point as *FMT*.

CU1 (1B): Customer Use 1

CU3 (3B): Customer Use 3

DC1 (11): Device Control Usually equipment-dependent. Same code point as *SBA*.

DC2 (12): Device Control Same code point as *EUA*.

DC3 (13): Device Control Same code point as *IC*.

DC4 (3C): Device Control Same code point as *RA*.

DEL (07): Delete Media-fill or time-fill character. Normally deleted on arrival at a receiving device or application.

DLE (10): Data Link Escape Used for transparent operation in binary synchronous communications (BSC) to indicate either:

- Start of transparent mode (when followed by *STX*)
- That the following byte is to be treated as a control character rather than as text
- That the following *DLE* character is to be treated as text

DS (20): Digit Select

DUP (1C): Duplicate 3270 format control order. It can be entered from the keyboard and is displayed as an overscored asterisk ($\overline{*}$). Its entry causes the modified data tag (MDT) bit in the field attribute byte to be set to 1. It is used to inform an application program that a duplicate operation, whose nature is usually application-dependent, is indicated for the rest of the field. Same code point as *IFS*.

Note—The character entered from the keyboard is, in fact, stored in 3270 buffer code form and is converted to the EBCDIC value, 1C, when it is transmitted to the host.

EM (19): End of Message 3270 format control order, used in an outbound data stream to a printer to mark the end of the data to be printed.

ENP (14): Enable Presentation SCS device-mode control code, used to enable the presentation, at the entering device, of device-entered data. Same code point as *RES*.

ENQ (2D): Enquiry Used in BSC point-to-point operation to bid for the link.

EO (FF) The highest-value EBCDIC code point. IBM defines EO as a 3270 format control order but does not assign a function to it.

EOT (37): End of Transmission Indicates the conclusion of a transmission, consisting of one or more messages.

ESC (27): Escape Used to indicate a special meaning or purpose for one or more subsequent characters.

ETB (26): End of Transmission Block Used to indicate the end of a block (frame) containing text that is continued in the next block.

ETX (03): End of Text Used to indicate the end of text in the last or only block of a message.

EUA (12): Erase Unprotected to Address 3270 outbound data stream order which stores nulls at all unprotected character locations up to, but not including, the stop address specified in the following two bytes. Same code point as *DC2*.

FF (0C): Form Feed 3270 format control order, used in printer operations to skip to the top of the next page.

FM (1E): Field Mark 3270 format control order. It can be entered from the keyboard and is displayed as an overscored semicolon ($\bar{;}$). Its entry causes the modified data tag (MDT) bit in the field attribute byte to be set to 1. It is used to inform an application program of the end of a field in an unformatted buffer or, more frequently, the end of a subfield in a formatted buffer. Same code point as *IRS*.

Note—The character entered from the keyboard is, in fact, stored in 3270 buffer code form and is converted to the EBCDIC value, 1E, when it is transmitted to the host.

FMT (2B): Format SCS data-defining control code, used with a 1-byte parameter to define the start of a formatted data stream. Same code point as *CSP*.

FS (22): Field Separator

GE (08): Graphic Escape 3270 inbound or outbound data stream order, used to introduce a graphic character from an alternate character set. The following byte contains the code point for the alternate graphic and has a value in the range 40 to FE (hexadecimal).

HT (05): Horizontal Tab Advances the active position to the next predetermined character position on the same line.

IC (13): Insert Cursor 3270 outbound data stream order which repositions the cursor to the current buffer address. Same code point as *DC3*.

IFS (1C): Interchange File Separator Used to indicate a logical boundary between files. Same code point as *DUP*.

IGS (1D): Interchange Group Separator Used to indicate a logical boundary between items called groups (e.g., groups of records within a file). Same code point as *SF*.

INP (24): Inhibit Presentation SCS device-mode control code, used to inhibit the presentation of keyboard-entered data at the entering device. Same code point as *BYP*.

IR (33): Index Return

IRS (1E): Interchange Record Separator Used to indicate a logical boundary between records. Same code point as *FM*.

IT (39): Indent Tab

ITB (1F): Intermediate Text Block Used to separate text (e.g., messages) within a BSC block (frame). Same code point as *IUS*.

IUS (1F): Interchange Unit Separator Used to indicate a logical boundary between units of information within a record or other larger item. Same code point as *ITB*.

LF (25): Line Feed Advances the active position to the same character position on the next line.

MF (2C): Modify Field 3270 outbound data stream order used to update Field Attributes and Extended Field Attributes at the current buffer address.

NAK (3D): Negative Acknowledgment Used to indicate the need to retransmit a block (frame).

NBS (36): Numeric Backspace Moves the active position backward by an amount equal to the width of a numeric character (0 to 9) on the same line.

NL (15): New Line 3270 format control order, used in printer operations to advance to the next line.

NUL (00): Null Has special meaning in 3270 operation. When 3270 presentation space data is sent to a host by means of a Read

Modified command, null characters are suppressed. Also, successful Insert-Mode entry of characters into a 3270 field depends on the existence of null characters in the field and to the right of the cursor. Displays as a space. The null code is 00 in both EBCDIC and 3270 Buffer Code.

POC (17): Program Operator Communication SCS control code which starts a 3-byte sequence, used to provide a communication mechanism between end users, where at least one of the end users is a terminal operator.

PP (34): Presentation Position SCS control code, used as the first of a 3-byte control sequence defining one of the following actions within a presentation space:

- An absolute or relative move
- A horizontal or vertical move
- A move and erase operation
- An erase to a new position, followed by a reset to the old position

PT (05): Program Tab 3270 outbound data stream order which advances the current buffer address to the first character position of the next unprotected field. Used in the transmission of data to an already-formatted display (presentation space). Same code point as *HT*.

RA (3C): Repeat to Address 3270 outbound data stream order used to store a specified character in all locations starting at the current buffer address and ending at the last location before a specified stop address. It is followed by a 2-byte stop address and the character to be repeated [which may be preceded by a *GE* (graphic escape) code]. Same code point as *DC4*.

RES (14): Restore Same code point as *ENP*.

RFF (3A): Required Form Feed

RNL (06): Required New Line

RPT (0A): Repeat

SA (28): Set Attribute 3270 inbound or outbound data stream order, used to specify an attribute type/value pair to be applied to subsequent characters until either:

- A new SA order (specifying the same attribute type) changes it
- Another write-type command is sent

- The operator presses the CLEAR key
- Power is turned off

Attribute types are highlight, color, and symbol set. Highlight values are for none, blink, inverse video, and underline; color values are for none, red, blue, green, pink, turquoise, yellow, and white; symbol set values are for base (0), APL/Text (1), or any of six programmed symbol sets.

SA orders are inserted in inbound data streams whenever there are changes from default (extended field attribute) values.

SBA (11): Set Buffer Address 3270 inbound and outbound data stream order, used to specify a new buffer address from which operations are to start or to continue. It is followed by a 2-byte buffer address. Same code point as *DC1*.

SBS (38): Subscript

SEL (04): Select SCS device control code, used with an associated 1-byte function parameter, to control a function within a device.

SF (1D): Start Field 3270 inbound and outbound data stream order, indicating the start of a field. Identifies the next byte as a field attribute byte. Same code point as *IGS*.

SFE (29): Start Field Extended 3270 inbound and outbound data stream order, used to specify field attribute and extended field attribute values for the field which follows. This code is followed by a count of the number of type/value pairs (see *SA*), then as many type/value pairs as the count specifies. The field attribute is specified by means of a type/value pair with a type of C0 (hexadecimal).

SI (0F): Shift In Used to indicate that subsequent character codes are to represent the graphic characters of the standard character set.

SM (2A): Set Mode Same code point as *SW*.

SO (0E): Shift Out Used to indicate that subsequent character codes are to represent a new set of graphic characters defined as an extension of the standard character set.

SOH (01): Start of Header Indicates that a message (frame) heading follows.

SOS (21): Start of Significance

SPS (09): Superscript

STX (02): Start of Text Indicates that text information follows.

SUB (3F): Substitute Used to replace a character that is known to be in error.

SW (2A): Switch Same code point as *SM*.

SYN (32): Synchronous idle Used in BSC practice to establish synchronization by byte. Normally used in pairs.

TRN (35): Transparent SCS data-defining control code, used to indicate the start of a transparent data stream. This code is always followed by a 1-byte binary count, indicating the number of bytes of transparent data.

UBS (1A): Unit Backspace

VT (0B): Vertical Tab Advances the active position to the same character position on a predetermined line.

WUS (23): Word Underscore

Section

6

Units of Measure (Système International)

Introduction

The correct use of units of measure in the computer and data communications field is an extension of the *Système International (SI)* of units of measure, the standard for a number of years for the scientific community.

SI defines *base units, supplementary units,* and *derived units.* The *base units* are *meter* for length, *kilogram* for mass, *second* for time, *ampère* for electric current, *kelvin* for thermodynamic temperature, *mole* for amount of substance, and *candela* for luminous intensity. The *supplementary units* are *radian* for plane angle and *steradian* for solid angle. First- and second-order *derived units* are derived from *base* and *supplementary units.* Some are handled by means of expressions (e.g., m/s—meters per second), some by special names [e.g., coulomb (C), which is equal to an ampère second (A.s)].

Although SI has a linguistic bias toward English and other languages with Latin roots, it is accepted by the scientific community in all countries, even those with a non-Roman alphabet.

Current practice in the computer and data communications field is inconsistent, confusing, and often ambiguous. For example, MB is commonly used to mean megabytes, while megabits per second is often written as MBS, MBPS, or Mbps. Even mbps (millibits per second?) shows up occasionally. Many computer store and mail-order advertisements use the colloquial "megs" to refer to megabytes. The list is endless.

This section is offered as a guide to overcoming the confusion through the adoption of an existing international standard. The rules are not those of the author but are taken from or derived from SI documentation.

Rules Affecting Style

The value associated with a unit of measure is always separated from the symbol or abbreviation used for the unit by a space. For example, 25 MHz, not 25MHz. Where an amount is used adjectivally, the space is customarily replaced by a hyphen. For example, 25-MHz processor. [Note that temperatures, when expressed in degrees Celsius (°C) rather than kelvin (K), are the one exception to the rule requiring a space. 25°C is correct; 25 °C is incorrect.]

Origins of SI Symbols and Abbreviations

Most *basic units* (and *derived units* with special names) use the lower-case initial character of the English word for the unit. In most cases, it is also the initial character of the French, Spanish, Italian, and, sometimes, the German word. Thus, gram is abbreviated to g, meter to m, and so on. Because the number of units of measure exceeds the capacity of the Roman alphabet, many units use abbreviations consisting of two or more letters (e.g., lm for lumen).

Some *basic units* and many *derived units* use an uppercase abbreviation of the name of a scientist historically associated with the discovery of a phenomenon associated with the unit. Thus, ampère (current) becomes A, gauss (magnetic field strength) becomes G, henry (inductance) becomes H, watt (power) becomes W. (Note that, when spelled out, they do not start with a capital letter, even though they are derived from proper names.)

Where the abbreviations for two or more units are derived from names with the same initial, all but one will use a two-letter abbreviation, of which the first letter is always uppercase and the second is always lowercase. Thus hertz (frequency, in cycles per second) becomes Hz, weber (magnetic flux) becomes Wb.

Some units use Greek letters. The best known is ohm (resistance), which is abbreviated to Ω (Greek letter omega) but is very often spelled out (i.e., not abbreviated).

Multipliers

Multipliers are always placed immediately to the left of the associated unit of measure and, except when necessary to avoid ambiguity, are almost always lowercase. Again, a space separates them from the associated value. Thus 10 milligrams becomes 10 mg, 5 kilometers becomes 5 km, 25 megahertz becomes 25 MHz (m = milli, M = mega). One multiplier, micro (for millionth), uses the Greek letter μ (mu).

Areas and Volumes

Prior to the establishment of the Système International, the use of units such as cc (for cubic centimeters) and sq.m (for square meters) was commonplace. SI replaces this practice with one that is more consistent and more elegant. For example, square meters becomes m^2 and cubic centimeters becomes ml (for milliliters).

Units Applicable to Computers and Data Communications

In addition to memory capacities, data rates, and so on, vendors and purchasers of computer and data communications equipment often need to specify electrical and environmental characteristics. Therefore, this section includes information on voltage, current, heat dissipation, etc.

Units Not Envisaged by the Système International

Units such as bit and byte (or octet) are arbitrary. SI does not specifically deal with arbitrary units and, thus, does not provide for their abbreviation. Thus, bits per second becomes bit/s. As bit is already an abbreviation (binary digit) and byte is a coined word and is deliberately short, this does not present a problem. The word "character" may be a problem; however, shortening it to four letters (as in char/s, for characters per second) does not introduce ambiguity.

Units for Data Communications Operating Variables and Equipment Specifications

The following table attempts to provide all the units which might be encountered in the specification or use of data communications equipment.

For some units, an attempt has been made to imply appropriateness. For example, the only unit indicated for the specification of cable lengths is the meter. It would be inappropriate, for example, to specify cable lengths in millimeters. First, it would involve large numbers. Second, it would imply a nonsensical degree of precision.

Note that, although the emphasis is on metric SI-derived units, some nonmetric measurements do persist, even in countries where the metric system is standard. Their form in the following table is, however, consistent with SI practice (e.g., char/inch, not cpi).

SI-Conforming Units for Data Communications Operating Variables and Equipment Specifications		
Measurement	**Unit**	**Abbrev.**
Air flow (cooling fan capacity, etc.)	liters per minute cubic meters per minute	l/min m^3/min
Altitude (maximum operating)	meter	m
Data rate, serial by bit	bits per second kilobits (10^3 bits) per second megabits (10^6 bits) per second	bit/s kbit/s Mbit/s
Data rate, serial by byte	bytes per second kilobytes (10^3 bytes) per second megabytes (10^6 bytes) per second	byte/s kbyte/s Mbyte/s
Data rate, serial by character	characters per second	char/s
Dimension—cable lengths, etc.	meter	m
Dimension—floor space, etc.	square meter	m^2
Dimension—floppy disks, **specific**	3½" (3.5 inch)—8.9 centimeters 5¼" (5.25 inch)—13.3 centimeters *[Use of metric (SI) units for disk diameters is rare, even in countries where metric units are the norm]*	8.9 cm 13.3 cm
Dimension—large components, surfaces, screen sizes, etc.	centimeter (10^{-2} m) square centimeter *(Display screen sizes are often specified in inches, even in countries where metric units are the norm)*	cm cm^2
Dimension—small components, tolerances, etc.	millimeter (10^{-3} m)	mm
Dimension—enclosure volume, etc.	liter cubic meter	l m^3

SI-Conforming Units for Data Communications Operating Variables and Equipment Specifications

Measurement	Unit	Abbrev.
Distance—data links, etc.	kilometer (10^3 m)	km
Electric current	ampère milliampère (10^{-3} A)	A mA
Electric potential ("voltage")	volt microvolt (10^{-6} V) *(signal level)* millivolt (10^{-3} V) kilovolt (10^3 V) *(applicable to display monitor power supplies)*	V µV mV kV
Electric power consumption (component level)	watt milliwatt (10^{-3} W) *Note: "active power"*	W mW
Electric power rating (power supply)	volt ampère *Note: Not watt, but a special case, in electric power technology, referred to as "apparent power."* kilovolt ampère (10^3 VA)	VA kVA
Electrical capacitance	microfarad (10^{-6} F) nanofarad (10^{-9} F) picofarad (10^{-12} F)	µF nF pF
Electrical resistance (DC) or impedance (AC)	ohm milliohm (10^{-3} Ω) kilohm (10^3 Ω) megohm (10^6 Ω)	Ω mΩ kΩ MΩ
Frequency	hertz (occurrences or cycles per second) kilohertz (10^3 Hz) megahertz (10^6 Hz) gigahertz (10^9 Hz)	Hz kHz MHz GHz
Heat dissipation	watt (joule per second—J/s) kilowatt (10^3 W)	W kW
Mass (**not** weight)	gram milligram (10^{-3} g) centigram (10^{-2} g) *(rare)* kilogram (10^3 g)	g mg cg kg
Memory capacity—chip	kilobit [1024 (2^{10}) bits] megabit [1048576 (2^{20}) bits]	kbit Mbit
Memory capacity—internal	byte kilobyte [1024 (2^{10}) bytes] megabyte [1048576 (2^{20}) bytes] gigabyte (2^{30} bytes) *In some contexts (e.g, CCITT Recommendations), **octet** is preferred to **byte**.*	byte kbyte Mbyte Gbyte

SI-Conforming Units for Data Communications Operating Variables and Equipment Specifications

Measurement	Unit	Abbrev.
Memory capacity—external (magnetic tape, mainframe direct-access devices, etc.)	kilobyte (10^3 bytes) megabyte (10^6 bytes) gigabyte (10^9 bytes) terabyte (10^{12} bytes)	kbyte Mbyte Gbyte Tbyte
Print density—horizontal and vertical	characters per inch *(not SI)* lines per inch *(not SI)* characters per centimeter lines per centimeter *[Use of metric (SI) measurements for print density is almost nonexistent, even in countries where metric units are the norm]*	char/inch *or* char/in line/inch *or* line/in char/cm line/cm
Printer speed	characters per second *(serial)* lines per minute *(line-by-line)*	char/s line/min
Speed—linear (tape transport mechanism, etc.)	centimeters per second meters per second	cm/s m/s
Temperature (operating and storage)	kelvin degrees Celsius *(The use of degrees Celsius is still very common and is considered acceptable. The kelvin scale uses the same interval as the Celsius scale but uses absolute zero as its base. Except in the United States, the Fahrenheit scale is obsolete.)*	K °C
Time	hour minute second millisecond (10^{-3} s) microsecond (10^{-6} s) nanosecond (10^{-9} s) picosecond (10^{-12} s)	h min s ms µs ns ps
Volume (enclosure size, etc.)	liter milliliter (10^{-3} l) centiliter (10^{-2} l) deciliter (10^{-1} l) *Although not strictly correct, L (uppercase) is sometimes used for liter. SI also specifies the optional use of a cursive l.*	l ml cl dl

Spelling

Note that *liter*, *meter*, and *gram* are American spellings. Canada and a number of other English-speaking countries use *litre, metre,* and *gram*. The United Kingdom uses *litre, metre,* and *gramme* (notwithstanding Fowler's preference, in *Modern English Usage*, for *gram*). The use of the SI abbreviations avoids these differences.

Conversion of Units of Length, Area, Volume, Mass, Temperature and Heat Dissipation to and from SI Values

Many U.S. and almost all non-U.S. manufacturers design their products to metric specifications. However, until either legislation or industry consensus brings the United States into conformity with the rest of the world, there will be a need to convert.

It makes good business sense for U.S. manufacturers, who wish to export their products, to provide metric specifications. Conversely, both U.S. and non-U.S. manufacturers, who wish to sell their products in the U.S., may want to allow for the reluctance (diminishing but still present) of many of their prospective customers to deal with the metric system.

The following table provides conversion factors in both directions. Note that, in converting, one must pay attention to precision. Specifically, one should avoid excessive precision, while observing necessary precision. If, for example, a cable can have a maximum length of 5000 feet, a converted value of 1500 m is sufficiently precise. The more precise 1524 m implies a degree of precision probably not contemplated in the establishment of the 5000-foot specification. For components, enclosures, and so on, converted values should imply neither more nor less precision than manufacturing tolerances allow. Operating, storage, and ambient temperatures should be expressed in whole degrees. For heat dissipation, two significant figures are usually close enough.

Like the SI table, this table is limited to those units that might be useful in specifying the characteristics and operating environment of data communications equipment. The nonmetric units are expressed in both SI and, parenthetically, in traditional form (where it differs from the SI form).

Conversion of Selected Nonmetric Units from and to SI Units

Metric (SI)	SI to Nonmetric: Precise Conversion	SI to Nonmetric: Typical Approximation	Nonmetric to SI: Precise Conversion	Nonmetric to SI: Typical Approximation	Non-metric
Length and Distance					
1/cm	2.5400/inch	2.54/inch	0.3937/cm	0.394/cm	1/inch (1 pi)
mm	0.03937 inch	0.04 inch	25.400 mm	25.4 mm	inch (″)
cm	0.3937 inch	0.39 inch	2.5400 cm	2.54 cm	inch (″)
m	3.281 foot	3.3 foot	0.3048 m	0.3 m	foot (′)
km	0.6211 mile	0.625 (5/8) mile	1.6099 km	1.6 km	mile (mi)
Area					
cm^2	0.1550 $inch^2$	0.16 $inch^2$	6.4516 cm^2	6.5 cm^2	$inch^2$ (sq.in)
m^2	10.7639 $foot^2$	11 $foot^2$	0.092903 m^2	0.09 m^2	$foot^2$ (sq.ft)
Volume					
ml	0.06102 $inch^3$	0.06 $inch^3$	16.3871 ml	16.4 ml	$inch^3$ (cu.in)
l	0.03531 $foot^3$	0.035 $foot^3$	28.317 l	28 l	$foot^3$ (cu.ft)
m^3	35.3145 $foot^3$	35 $foot^3$	0.028317 m^3	0.03 m^3	$foot^3$ (cu.ft)
Mass					
g	0.0352740 oz	0.035 oz	28.3496 g	28 g	ounce
kg	2.20462 lb	2.2 lb	0.453593 kg	0.45 kg	pound
Thermal (temperature and heat dissipation)					
p°C	$p = \frac{5(q-32)}{9}$	$p = \frac{5(q-32)}{9}$	q = 1.8p+32	q = 1.8p+32	q°F
W	0.2932 BTU/h	0.3 BTU/h	3.410 W	3.4 W	BTU/h

Bibliography

Bartee, Thomas C., *ISDN, DECnet and SNA Communications,* Howard W. Sams & Company, Indianapolis, IN, 1989. ISBN 0-672-22512-3.

Black, Uyless, *The V Series Recommendations: Protocols for Data Communications over the Telephone Network,* McGraw-Hill, New York, NY, 1991. ISBN 0-07-005552-1.

Black, Uyless, *The X Series Recommendations: Protocols for Data Communications Networks,* McGraw-Hill, New York, NY, 1991. ISBN 0-07-005546-7.

CCITT, Blue Book, Volume III (fascicles 7, 8, and 9), *I-Series Recommendations.* ISBN 92-61-03371-7, 92-61-03381-4, and 92-61-03391-1.

CCITT, Blue Book, Volume VIII (all eight fascicles), *V-Series and X-Series Recommendations.* ISBN 92-61-03671-6, 92-61-03661-9, 92-61-03681-3, 92-61-03691-0, 92-61-03701-1, 92-61-03711-9, 92-61-03721-6, and 92-61-03731-3.

Comer, Douglas, *Internetworking with TCP/IP—Principles, Protocols and Architectures,* Prentice-Hall, Englewood Cliffs, NJ, 1988. ISBN 0-13-470154-2.

Cypser, T. J., *Communications Architecture for Distributed Systems,* Addison Wesley, Reading, MA, 1978. ISBN 0-201-14458-12.

Doll, Dixon R., *Data Communications — Facilities, Networks and Systems Design,* John Wiley & Sons, New York, NY, 1978. ISBN 0-471-21768-9.

IBM, *Fachausdrücke der Informationsverarbeitung, Wörterbuch und Glossar,* IBM Deutschland GmbH, 1985. IBM SQ12-1044-1.

IBM Systems Reference Library, *IBM 3174 Establishment Controller — Functional Description.* IBM GA23-0218-07.

IBM Systems Reference Library, *IBM 5250 Information Display System — Functions Reference Manual.* IBM SA21-9247-6.

IBM Systems Reference Library, *Service for Consultants* and *Universal Sales Manual.*

Kessler, Gary C., *ISDN Concepts Facilities and Services,* McGraw-Hill, New York, NY, 1990. ISBN 0-07-034242-3

McNamara, John E., *Technical Aspects of Data Communication,* Digital Press, 1988. ISBN 1-55558-007-6.

Martin, James, and Leben, Joe, Arben Group, *Data Communication Technology,* Prentice-Hall, Englewood Cliffs, NJ, 1988. ISBN 0-13-196643-X.

Martin, James and Leben, Joe, Arben Group, *Principles of Data Communication,* Prentice-Hall, Englewood Cliffs, NJ, 1972 & 1988. ISBN 0-13-709891-X.

Martin, James, and Chapman, Kathleen Kavanagh, Arben Group, *SNA: IBM's Networking Solution,* Prentice-Hall, Englewood Cliffs, NJ, 1987. ISBN 0-13-815143-1.

Sherman, Ken, *Data Communications—A User's Guide,* Prentice-Hall, Englewood Cliffs, NJ, 1981, 1985, & 1990. ISBN 0-13-199092-6.

Index

ABOUT THE AUTHOR

William F. Potts is president of WFP Consulting in San Jose, Calif. Previously, he was director of International Market Development, Communications Products, at Novell, Inc. Mr. Potts has held positions in data communications (for almost 16 of his 33 years in the computer field), in product management, and in sales and marketing management. Recently, he has presented data communications seminars in over a dozen countries. Mr. Potts has been a fellow of the British Computer Society since 1969.